Caring for Plastics in Collections

AF701832

Caring for Plastics in Collections presents basic knowledge of plastics in an accessible way and empowers readers to determine the best approach for care, taking into account the resources at their institution and the needs of their specific collection.

Drawing on the author's 20 years of experience working as an Objects Conservator, the book defines what a plastic is; which plastics are most worrisome in collections and why; and how you can identify plastics, monitor them, and mitigate issues. Clarifying that there are guidelines that all plastics will benefit from, Coughlin outlines a gradual, consistent approach to care that is supplemented by 25 case studies from 10 institutions. Knowing that collection teams are too frequently understaffed and funds are often limited, the book approaches the care of plastics from both preventive conservation and collection management with a focus on low-cost, accessible options for monitoring potential degradation products; the use of readily available materials and resources; tips for designing condition surveys; and using existing database content to flag plastics that may pose a collection risk. In this way, collection stewards, such as collection managers, registrars, curators, archivists, and anyone tasked with caring for collections containing plastics, will be poised to make more informed decisions about storage, display, and loans.

Caring for Plastics in Collections is an essential resource for professionals working in a range of institutions around the globe, including galleries, libraries, archives, historic houses, and museums.

Mary Coughlin is Associate Professor in Museum Studies at The George Washington University in the United States.

Caring for Plastics in Collections

Mary Coughlin

Routledge
Taylor & Francis Group
LONDON AND NEW YORK

Designed cover image: © Mary Coughlin

First published 2025
by Routledge
4 Park Square, Milton Park, Abingdon, Oxon OX14 4RN

and by Routledge
605 Third Avenue, New York, NY 10158

Routledge is an imprint of the Taylor & Francis Group, an informa business

© 2025 Mary Coughlin

The right of Mary Coughlin to be identified as author of this work has been asserted in accordance with sections 77 and 78 of the Copyright, Designs and Patents Act 1988.

All rights reserved. No part of this book may be reprinted or reproduced or utilised in any form or by any electronic, mechanical, or other means, now known or hereafter invented, including photocopying and recording, or in any information storage or retrieval system, without permission in writing from the publishers.

Trademark notice: Product or corporate names may be trademarks or registered trademarks, and are used only for identification and explanation without intent to infringe.

British Library Cataloguing-in-Publication Data
A catalogue record for this book is available from the British Library

ISBN: 9781032757131 (hbk)
ISBN: 9781032757124 (pbk)
ISBN: 9781003489658 (ebk)

DOI: 10.4324/9781003489658

Typeset in Times New Roman
by Newgen Publishing UK

To Bruno Pouliot and Beth Richwine for encouraging me to study plastics and for seeing my ability to teach long before I ever would have. I hope you knew how grateful I was to have you as mentors.

Contents

List of Figures *viii*
Sidebar Contributors *xiii*
Acknowledgments *xv*

1 Introduction 1

2 What Is Plastic? 6

3 Which Plastic? Clues to Identification 27

4 Problem Plastics 46

5 Influencers on Plastics: Environment 69

6 Influencers on Plastics: Material Choices 94

7 Low-Cost, On-Site Monitoring of Plastics 114

8 Plastics Collection Management Part 1: Processes and Procedures 130

9 Plastics Collection Management Part 2: Discussions and Decisions 155

10 Collection Care Approaches for All Plastics 169

Index *182*

Figures

2.1 An FDM printer creating a plastic object by depositing filament layer-by-layer. Credit: Jaemi Rowe, Rochester Institute of Technology 11

2.2 Lumarith salesman's kit of 49 CA samples on a metal chain. NMAH 1983.0538.61. Photographed by author, Division of Medicine and Science, National Museum of American History, Smithsonian Institution 21

2.3 Severely deteriorated CA Lumarith coupon (a) alongside intact neighbor (b). Note green corrosion of metal chain. NMAH 1983.0538.61. Photographed by author, Division of Medicine and Science, National Museum of American History, Smithsonian Institution 22

2.4 Rigid white plastic yellowed from exposure to sunlight and continued darkening after five years stored in a cabinet 23

2.5 Different coloring in the two-part handle of a toy teapot became evident after a decade 24

3.1 Positive reaction of diphenylamine on polyethylene bag used to store degrading cellulose nitrate 28

3.2 Positive reaction of Nitrate Test paper with plum colored "Celluloid" cellulose nitrate rod from 1941. NMAH 1983.0538.91. Photographed by author, Division of Medicine and Science, National Museum of American History, Smithsonian Institution 29

3.3 The Celluloid Club of Newark, NJ invitation, 1900. NMAH CH.334327. Photographed by author, Division of Medicine and Science, National Museum of American History, Smithsonian Institution 34

3.4 Bottom of Tupperware cup marked "Tupper Corporation Farnumsville, Mass. U.S.A. 122-2". NMAH 1985.3014.06. Photographed by author, Division of Home and Community Life, National Museum of American History, Smithsonian Institution 36

3.5 Polypropylene box (marked with RIC #5, PP) propped against a blank Word document with polarized sunglasses in front producing bright birefringence colors 39
3.6 Timeline for dates when plastic was broadly manufactured, not invented. Biodegradable plastic date is for discovery (Barrett 2018). Dates may slightly differ depending on resource (MoDiP (Museum of Design in Plastics) 2007b; Plastic Identification Tool n.d.; van Oosten 2022) 43
4.1 Discoloration, crazing, and embrittlement in thick CN sample from 1942. NMAH 1983.0538.87. Photographed by author, Division of Medicine and Science, National Museum of American History, Smithsonian Institution 51
4.2 Circa 1930s geometric model made of a CN internal form and CA external form. CA is yellowed, weeping, and warped from plasticizer migration. Once in the NMAH mathematics collections (1979.0102.058). Photographed by author, Division of Medicine and Science, National Museum of American History, Smithsonian Institution 53
4.3 Circa 1930s geometric model made of a CN internal form and CA external form. CA is yellowed and warped; curled section is rigid from plasticizer loss, some of which crystallized. Once in the NMAH mathematics collections (1979.0102.312). Photographed by author, Division of Medicine and Science, National Museum of American History, Smithsonian Institution 54
4.4 Fume fading of acetate fabric caused by contact with CN buttons 55
4.5 Mockup of buttons with silicone release paper barrier. Credit: Christopher Mazza 56
4.6 Swatch watch with tacky, yellowed band and watch face mount. Stored flat in open storage. NMAH 1996.0089.07. Photographed by author, Division of Work and Industry, National Museum of American History, Smithsonian Institution 60
4.7 Swatch watch with yellowed band and watch face mount. Stored curved in sealed rigid plastic container 60
4.8 Polyurethane foam upholstery replacement during treatment of an Alexander Girard chair for Braniff Airways, treated by Kress Fellow Geena Roth. Note darker discolored surfaces in (a) that are crumbling into powder (b). Credit: Jessica Walthew for Cooper Hewitt Smithsonian Design Museum 63
5.1 Principle of reciprocity for light exposure 73
5.2 Joseph Beuys, *Postcards 1968–1974*, 1974, wood, PVC, and cardboard, $18 \times 12.5 \times 6\ cm^3$. Harvard Art Museums/Busch-Reisinger Museum, 1995.295. PVC component rehoused in Mylar to prevent the loss of plasticizer. Image courtesy of Harvard Art Museums 78

5.3 Circa 1930s geometric model made of a CN internal form and CA external form. The leached plasticizers from the CA crystallized at lower RH. Once in the NMAH mathematics collections (1979.0102.058). Photographed by author, Division of Medicine and Science, National Museum of American History, Smithsonian Institution 81
5.4 Isotherm demonstrating the relationship among temperature, equilibrium relative humidity, and equilibrium moisture content for a cellulose triacetate film plasticized with 8% diethyl phthalate and 8% triphenyl phosphate (wt%) 84
5.5 (1a) Moisture-induced expansion (strain) at 20°C for different cellulose triacetate film plasticized and non-plasticized, and with and without a gum-arabic-based paint layer at indicated levels of RH; (1b) calculated hygroscopic expansion coefficient representing the change in strain per percentage change in relative humidity at indicated transition steps 87
6.1 Older yellowed LDPE bags alongside newer LDPE bags to store small objects 99
7.1 Monitoring severely degraded CN samples with silver, lead, and copper coupons over four weeks. As a strong oxidizing agent, the nitric acid corroded the copper after just one week. The bright pink in the image is reacted Cresol Purple indicators seen in Figure 7.3. NMAH 1983.0538.87/.90/.91. Photographed by author, Division of Medicine and Science, National Museum of American History, Smithsonian Institution 116
7.2 Celanese Celluloid Corp. CN color sample from 1944 showing uneven aging, including level of acidic off-gassing after three days of monitoring with A-D Strips. NMAH 1983.0538.93. Photographed by author, Division of Medicine and Science, National Museum of American History, Smithsonian Institution 117
7.3 (a) Cresol purple with CN color samples from the early 1940s, (b) Reaction after a few minutes, (c) Reaction after a few hours. The metal grommets of the storage box are heavily corroded and the paper box itself is an acidic pH 3. NMAH 1983.0538.87/.90/.91. Photographed by author, Division of Medicine and Science, National Museum of American History, Smithsonian Institution 123
7.4 Eyeglasses monitored for three days. A-D Strip turned light green (positive reaction, compared to unused blue control strip). Cresol purple stayed yellow (no reaction). Monitoring helped confirm glasses as CA, not CN. NMAH MG.314637.12189. Photographed by author, Division of Medicine and Science, National Museum of American History, Smithsonian Institution 124
7.5 Eyeglasses monitored for three days. A-D Strip turned light green (positive reaction, compared to unused blue control strip). Cresol

purple turned pink (positive reaction). Monitoring helped confirm glasses as CN, not CA. NMAH 1984.0140.383. Photographed by author, Division of Medicine and Science, National Museum of American History, Smithsonian Institution 124
7.6 FTIR spectrum of PVC with polymer degraded and significant loss of plasticizer. Credit: Stephen Boyes, Associate Professor, Chemistry, George Washington University 126
7.7 FTIR spectrum of PVC with polymer in good condition and plasticizer intact. Credit: Stephen Boyes, Associate Professor, Chemistry, George Washington University 127
7.8 Monitoring degrading PVC goggles with A-D Strip, Ozone Test strip, and copper coupon for 30 days along with control references. Positive reactions of the A-D Strip and Ozone Test stick support FTIR results that the PVC is severely degraded. No reaction of copper coupon 128
8.1 Condition, Current Housing, and Conservation Priority pie charts showing proportions of the collection needing attention for housing and treatment, and the prioritization recommendations. Credit: Kate Wight Tyler 141
8.2 Database entry containing the categories and respective signs of degradation that can be assigned to each individual plastic part of the object ("Plastic parts") 145
8.3 Boots (C.2014.480.1) worn by Carol Orazem on September 11, 2001. (a) as they appeared upon acquisition in 2014 and (b) in 2024 in plastic enclosures, note severe deterioration of the soles. Credit: Collection 9/11 Memorial & Museum, Gift of Carol Orazem-DiPaolo 148
8.4 Catalog number applied to CN by printing on paper and coating with an acetone lacquer. Solvent interaction caused blistering, making the label illegible. Deaccessioned. Photographed by author, Division of Medicine and Science, National Museum of American History, Smithsonian Institution 150
9.1 Spock ear tips from *Star Trek II: The Wrath of Kahn* photographed in 2005 as part of a denied loan request. NMAH 1999.0102.01/.02. Photographed by author, Division of Culture and the Arts, National Museum of American History, Smithsonian Institution 159
9.2 Steven O'Banion and Shervone Neckles in the artist's studio during the interview in front of elements from *Domiciliation: Bless This House* and other works. Photographed by Samantha Owens 163
9.3 Circa 1930s geometric model made of a CN that is now deteriorated with scattered tape residue from previous repairs. Once in the NMAH mathematics collections

(NMAH 1979.0102.254). Photographed by author, Division of Medicine and Science, National Museum of American History, Smithsonian Institution 167

10.1 Lumarith salesman's kit sample coupon. (a) Front in visible light, (b) Front in UV light (365 nm), (c) Back in UV light (365 nm). Without UV examination, cracks were obscured by weeping materials. NMAH 1983.0538.61. Photographed by author, Division of Medicine and Science, National Museum of American History, Smithsonian Institution 170

10.2 CN sample in (a) visible light, (b) UV light (365 nm). NMAH 1983.0538.83. Photographed by author, Division of Medicine and Science, National Museum of American History, Smithsonian Institution 171

10.3 (a) Silicone rubber spatula in 2011; orange weeping material pooled at bottom and dust stuck to the surface. (b) Spatula in 2021; orange weeping material dried, surface darkened, and dust stuck to surface. NMAH 2001.0253.0707. Photographed by author, Division of Home and Community Life, National Museum of American History, Smithsonian Institution 177

10.4 Dishwashing gloves in 2021. Top black glove was blue in 2011 but turned completely black ten years later. The bottom glove was protected and retained most of its blue color. Photographed by author, Division of Home and Community Life, National Museum of American History, Smithsonian Institution 178

Sidebar Contributors

Sarah Barack, Head of Conservation and Senior Objects Conservator, Cooper Hewitt, Smithsonian Design Museum

Kris Cnossen, Textile and Costume Conservator, Midwest Textile Conservation, LLC

Susan Costello, Associate Conservator of Objects and Sculpture, Harvard Art Museums

Kiri Douglas, Andrew W. Mellon Fellow in Paper Conservation, National Gallery of Art

Wendi Field Murray, PhD, Archaeology Collections Manager, Wesleyan University

Kate Fugett, Preventive Conservator, National September 11 Memorial & Museum

Elena Gómez-Sánchez, PhD, Scientist, Deutsches Bergbau-Museum Bochum, Leibniz-Museum für Geo-Ressourcen

Kerith Koss Schrager, Head of Conservation, National September 11 Memorial & Museum

Till Krieg, Objects Conservator, Kultur- und Stadthistorisches Museum Duisburg

Nicole Ledoux, Associate Conservator of Objects and Sculpture, Straus Center for Conservation and Technical Studies, Harvard Art Museums

Christopher Mazza, Associate Conservator, The Costume Institute, The Metropolitan Museum of Art

Cristian Mazzon, PhD, Conservation Scientist, Deutsches Bergbau-Museum Bochum, Leibniz-Museum für Geo-Ressource

Steven O'Banion, Director of Conservation, Glenstone Museum and Voices in Contemporary Art (VoCA) Board Member

Jacek Olender, PhD, Research Fellow, Image Permanence Institute, Rochester Institute of Technology

Samantha Owens, Associate Conservator, Glenstone Museum

Georgina Rayner, PhD, Conservation Scientist, Straus Center for Conservation and Technical Studies, Harvard Art Museums

Emma J. Richardson, PhD, Director of Research, Image Permanence Institute, Rochester Institute of Technology

Meredith Sharps Noyes, PhD, Research Scientist, Image Permanence Institute, Rochester Institute of Technology

Julie Unruh, Objects Conservator, Julie Unruh, LLC

Jessica Walthew, Conservator, Cooper Hewitt, Smithsonian Design Museum

Kate Wight Tyler, Conservator, Brooklyn Museum of Art

Acknowledgments

Sophie Rowe ACR, Head of Conservation, Fitzwilliam Museum, University of Cambridge, for her enthusiastic support of this volume and for reviewing each chapter.

Curators Steve Velasquez and Paula Johnson at the Smithsonian Institution's National Museum of American History for inviting students to work in Julia Child's kitchen, welcoming me back over the years to check on the plastics, and approving the kitchen case study.

Shervone Neckles for discussing the use of plastics in her artwork during an interview that Steven O'Banion and Samantha Owens from Glenstone Museum specifically arranged for this book.

The Museum Studies students of The George Washington University and the Conservation students and instructors at Winterthur/University of Delaware and New York University – the plastics sessions are always the highlight of my academic year, thanks for letting me believe they are for you too!

1 Introduction

The primary goal of this book is to make knowledge about plastics accessible and guide the application of that knowledge into practical approaches for collection care for diverse holdings, including art, design, history, natural history, and special collections. I anticipate that readers will have a range of job titles, including collections manager, registrar, curator, archivist, or conservator, and work at any number of collecting institutions like museums, galleries, historic sites, archives, and libraries. Regardless of job title or collection type, a commonality in the heritage sector is the presence of plastics. Rather than being uniform, the term plastic actually encompasses a variety of forms and formulations that have different risks and sensitivities. Whether as accessioned objects or as materials used for storage, display, or transport, plastics are everywhere, and an understanding of them is critical for responsible stewardship.

There are stellar resources in print and online about plastics in collections, many referenced throughout this book, but sometimes their focus on chemistry can be intimidating and, therefore, may lose a critical audience. Key terms and concepts for understanding properties of various plastic types, degradation mechanisms, and the impact that environmental and material choices can have are discussed so that if non-specialists want to pursue more technical resources, a foundational understanding is established. By becoming familiar with the properties of plastics that impact function, aging, and risks, more informed collection care decisions can be made at all institutions regardless of staff size or budget.

As a Professor in a Museum Studies graduate program, I use the term "museum" throughout the book as a catchall for collecting institutions. So, it is important to remember that the content being discussed, whether about plastics as a material, aspects of preventive care, or collection management decisions, applies to all collections regardless of the designation of the entity that houses them.

As an Objects Conservator, I default to using the term "object," but it is meant in the broadest sense. So, in this volume, a plastic object is anything three-dimensional, such as an artwork, postcard, sculpture, kitchen utensil, eyeglasses, jewelry, or toy. That said, though intended as a broad term, there are times when using "object" is not appropriate, so more specifics are used when needed. As caretakers of diverse collection holdings, I did not want to assume that your responsibilities are limited

DOI: 10.4324/9781003489658-1

to three-dimensional plastics, so I solicited input from a paper conservator on synthetic paper and two textile conservators, one in private practice and the other from The Costume Institute at The Metropolitan Museum of Art, to provide insight on different aspects of human-made fibers.

In order to bound the content, the focus of the book is on semi-synthetic and fully synthetic plastics, primarily as three-dimensional objects. As a result, natural plastics like rubber are only briefly mentioned, but there is discussion about the emergence of biobased plastics. Synthetic paints like latex and acrylic are not included, nor are coatings and lacquers discussed. Film formats, namely cellulose nitrate and cellulose acetate, are touched upon, but with the wealth of readily available information on degradation and care available, plastic-based films are not covered in-depth.

Book Sections

The book is organized into three main sections:

- Plastic as a material
- Preventive conservation
- Collection management decisions and actions

Chapters 2–4 focus on plastic as a material. Chapter 2 defines characteristics of plastics to help answer, "What is plastic?". Chapter 3 describes options for identifying plastics, including descriptions of the most commonly applied methods of scientific analysis, using both analytical instruments as well as lower-tech options. This is followed by more accessible identification methods that do not rely on access to instrumentation or a laboratory but rather use clues from appearance and curatorial information like the dates of use or when accessioned. By providing a range of options for identifying plastics, or at least helping to narrow down to a probable type, potential problems can be flagged and more tailored approaches applied to care. The five plastics that are known to be problematic in collections comprise Chapter 4.

The next three chapters (Chapters 5–7) are focused on preventive conservation. Chapter 5 covers the influence of the surrounding environment (light, temperature, relative humidity, and pollutants) on how plastics age. Chapter 6 pivots from plastics being accessioned objects to plastics as materials used for storage, display, and transport and the potential impact of those decisions. The final chapter in the preventive conservation section (Chapter 7) outlines low-cost, on-site ways to monitor condition issues that plastics may develop over time.

After Chapter 7, the focus transitions to collection management decisions, which are split between two chapters. Chapter 8 discusses processes and procedures for:

- Handling
- Documentation
- Databases
- Labeling

Chapter 9 considers issues around the discussions that go into decision-making about:

- Accessioning
- Exhibitions and Loans
- Replacement or Replication
- Deaccessioning

Chapter 10 presents a nine-step approach that draws together the key preventive conservation and collection management factors detailed in the book and illustrates how they can be applied to any collection so you will be empowered to systematically assess the needs of the plastics under your care. The first seven steps cover actionable items of observation, confirming the observation, deciding if you need to separate and isolate plastics, cleaning the area degrading plastics were kept in, evaluating support materials assessing the environment to determine if changes are feasible and beneficial, and the importance of documentation for plastics. The last two steps encourage discussion about whether replacement or replication could help fulfill the maker's intent or the museum's mission or if deaccessioning the plastic is a prudent option. Julia Child's Kitchen, on display at the Smithsonian's National Museum of American History, serves as a case study to illustrate this step-wise approach.

Chapter Features

Each of the ten chapters opens with a brief introduction and then is subdivided with headings to organize subsections within the chapter. The intention of each chapter is to highlight approaches and practices as they apply to plastics in collections with sources cited from books, peer-reviewed journals, and online resources. Figures, including tables, images, and graphs, are placed throughout the book to illustrate concepts. Each chapter has at least one small case study, referred to as a sidebar. In total, there are 25 sidebars:

Chapter 2:

- Synthetic Paper
- 3D Printed Plastics
- Textile Additives
- Potential Health Impacts and Precautions for Handling Plastics

Chapter 3:

- Dating Tupperware
- Polarizing Filters and Birefringence

Chapter 4:

- Cellulose Nitrate and Cellulose Acetate Buttons
- Comparison of Storage Impact on Plasticized PVC Watches

Chapter 5:

- The Positive Results of Rehousing a Flexible Polyvinyl Chloride Object
- Physical Response of Plastics to the Environment: Equilibrium Moisture Content
- Physical Response of Plastics to the Environment: Humidity Expansion Coefficient
- Anoxic Storage

Chapter 6:

- Storage Plastics and Sample Viability in Archaeological Research

Chapter 7:

- Monitoring a Cellulose Acetate Necklace

Chapter 8:

- Handling Human-Made Fiber-Based Textiles
- Artist Interview
- Identifying and Tracking Color Change
- Designing Plastics Surveys for Successful Follow-up
- Tools for Surveying Plastic Collections: Monitoring Aging Phenomena
- Collections Survey Planning at the 9/11 Memorial & Museum
- Labeling Textiles

Chapter 9:

- Loan Request of Deteriorating Plastic
- An Artist Interview Case Study Highlighting the Significance of Plastics
- Deaccessioning Deteriorating Plastics

Chapter 10:

- Plastics in Julia Child's Kitchen

Sidebar contributions are by conservators, conservation scientists, curators, and collections managers with experience in a range of collections, including those focused-on art, history, design, and archaeological materials from the following institutions:

- Cooper Hewitt, Smithsonian Design Museum
- Deutsches Bergbau-Museum Bochum, Leibniz-Museum für Geo-Ressourcen (German Mining Museum)
- Glenstone Museum
- Harvard Art Museums
- Image Permanence Institute
- National Gallery of Art
- National September 11 Memorial & Museum
- Smithsonian National Museum of American History
- The Costume Institute, The Metropolitan Museum of Art
- Wesleyan University

Whether in a museum, gallery, historic site, archives, or library, there are likely a myriad of plastics accessioned into the collection with others found in the products used for storage, display, and transport. The variety of plastics and the uncertainty that can come with them can make setting up a systematic approach for determining needs overwhelming. Through accessible language, real-world examples, and practical insights, this book aims to serve as a starting guide for caring for plastics in collections.

2 What Is Plastic?

What is plastic? This seems like a straightforward question. We can all picture a plastic but we all won't picture the same type of plastic. For some, the imagined plastic may be a toy like a doll, frisbee, or ball. For others it may be an imitation leather handbag, photograph, sculpture, jewelry, or something commonplace like a soda bottle or disposable cutlery. And even if two people name the same item, the look of the envisioned plastic will likely differ by color or size because the versatility of plastic allows it to take form into almost anything we can imagine.

Now take a moment to consider the plastics you touch during a day from when you wake up, go to work or run errands, sit down to dinner, and get ready for bed. How many different forms do these plastics take? Which are rigid? Perhaps these are your smartphone, car steering wheel, or eyeglasses. Are any flexible like contact lenses? Which are in sheet form like a shower curtain? Are some forms cushioned with foam? Which are stretchy? Take a look at the label on your clothing; is Spandex or Lycra listed? If so, you are wearing plastic. What colors are these plastics? How many are transparent? Which are utilitarian and which provide frivolous joy? Which could you imagine being accessioned into a museum? Likely every plastic you came into contact with during your day is in a museum already! Plastic's adaptability, usefulness, and low cost have made it integral to material culture for well over 100 years and are found in every museum collection whether as an accessioned object or as a material used in collection care.

Plastic Categories

Plastics can be grouped into three categories based on the materials they are primarily derived from:

- Natural
- Semi-synthetic
- Synthetic

Natural plastics like horn and tortoiseshell can be modified into new shapes with heat, moisture, and pressure. Natural rubber can be modified through

DOI: 10.4324/9781003489658-2

Table 2.1 Common Human-Made Fibers

Common Human-Made Fibers	
Semi-synthetic *Made by extracting polymer building blocks from a natural material*	Synthetic *Often made from petroleum or natural gas*
Cellulose-based: rayon and acetate (triacetate)	Nylon (6 and 6,6)
Mineral-based: glass, ceramic, metal, and carbon fibers	Polyester
Protein-based: Azlon	Acrylic
	Polyurethane: Spandex
	Aramid: Kevlar and Nomex

Sources: Quye (2014); Dress and Textile Specialists and Plastics Subject Specialists Network (2022).

Credit: Kris Cnossen.

vulcanization, a curing process that bonds rubber and sulfur (see Elastomers below and Chapter 4). While natural plastics will not be covered in this volume, it is important to note their classification here.

In the 19th century, people's desire for the appearance of luxury goods at lower costs contributed to the success of **semi-synthetic** plastics whose mixture of natural and human-made materials could mimic elite products like tortoiseshell and ivory. Eventually, semi-synthetic plastics expanded beyond imitating nature. To make semi-synthetic plastics, chemists modify a natural polymer, such as cellulose, in order to achieve the desired form which could be a transparent film, three-dimensional object, or textile fiber all available in just about any conceivable color. Prime examples in this category for textiles include rayon and acetate (Table 2.1). Cellulose nitrate and cellulose acetate are early semi-synthetics that are in many museum collections as film and three-dimensional objects (Chapter 4).

In the early and mid-20th century, chemists developed fully **synthetic** plastics and with these formulations come a wider variety of forms and functions. The plastics in this category are often easy to identify because their names usually start with "poly" to denote their polymer type (more on polymers below). An important defining element of synthetic plastics is that they are mainly derived from non-renewable fossil fuels like petroleum, natural gas, and coal with the formulation of 98% of plastics today being based on these sources (van Oosten 2022 and Baker 2023). For textiles, the creation of synthetic fibers, like nylon, polyester, acrylic, and polyurethane (PUR), helped broaden options for designers and consumers through improved elasticity, decreased wrinkling, and more color options (Table 2.1). PUR and polyvinyl chloride (often referred to by its abbreviation PVC) can also be used to imitate leather.

Sidebar 2.1 Synthetic Paper

Kiri Douglas

Synthetic paper, as a category, is ambiguously defined. The word "paper" is simply used to represent the end use of this type of product. The term synthetic paper encompasses a group of substrates that are paper-like but composed of plastic rather than cellulose. Synthetic paper is prevalent in everyday life as labels, menus, envelopes, and banknotes. There are three major categories of synthetic papers based on manufacturing technique:

- Synthetic fiber and filaments
- Extruded film
- Hybrid

Synthetic fiber-based substrates can be produced through various nonwoven processes, such as spunmelt, airlaid, drylaid, or wetlaid. The latter utilizes conventional papermaking equipment. Extruded films can be produced via flat film or blown film methods. Stone Paper® is one example of an extruded film with a high calcium carbonate content (>70%) which produces a mineral rich synthetic paper (Lunk and Stange 1971; Mustalish 2007). Hybrid configurations include conventional cellulosic papers laminated with plastic films or with plastic coatings.

The demand for new specialty and technical papers for medical, pharmaceutical, and horticultural industries, among others, has spurred the rapid technological advancements in the use of synthetic fibers for papermaking. Substrates that can perform operational functions while being resilient across a variety of environmental conditions and stressors, such as solvent exposure, bacteria, and extreme temperatures, have required the application of plastics within the papermaking process. Arledter states that it is the "active and close cooperation between the papermaker, the non-woven technologist, the synthetic fiber producer, the resin industry and the machine builder" that has advanced the web bonding process since the 1960s (1964, 6). The paperization of plastic films has further increased synthetic paper's effectiveness in challenging environments.

Synthetic paper began to be available on the retail market from the mid-1960s, with utilitarian papers like Kimdura (based on an extruded film) used experimentally by artists such as Lita Albuquerque. Papers produced specifically for the art market, such as 3M Art Fabric (based on synthetic fibers), were used by Corita Kent and Sam Gilliam. Since then, a number of synthetic paper brands have increased, including those advertised directly to artists, such as Yupo, Strathmore Aquarius II, and the aforementioned Stone Paper®. These are most commonly found within collections as paper clothing, maps, book covers, and artworks.

While synthetic paper does not deteriorate by the same chemical mechanisms as cellulosic paper, they do yellow as they age due to several factors that depend on the kind of plastic binders and additives used within the paper. High heat and light, especially ultraviolet (UV), expedites the deterioration of the polymers, eventually causing brittleness and fracturing of the substrate (Chu and Nel 2019). Caring for these materials within collections should involve periodic observation to look for changes, low light levels, UV-resistant glazing, digitization or migration of information, and maintaining a stable climate. UV imaging of synthetic paper based on extruded films shows fluorescence from fingerprints; therefore, nitrile gloves should be worn when handling. To date, synthetic papers have had limited research; further study will help strengthen collection care strategies.

Growing sustainability efforts and the desire to move away from fossil fuel-based non-renewable resources have led to the growth of biobased and biodegradable synthetic plastics. As the name implies, biobased plastics have their basis in natural materials like corn or sugarcane. An example of this that is growing in general use, as well as entering more museum collections, is polylactic acid (PLA) which is often used in 3D printing (Sidebar 2.2). Chemists can now create many plastics including polyethylene (PE), polypropylene (PP), polyethylene terephthalate (PET), and PUR from biobased rather than petroleum-based sources. Fossil fuel-based plastics and those from plants have the same polymer structure; the only way to distinguish between them is through radiocarbon dating since fossil fuels are older than plants (Baker 2023). It is important to realize that just because biobased plastics come from renewable sources, it does not mean that they are biodegradable or compostable and their additives can be just as harmful (Sidebar 2.4). How biobased plastics will age in museum environments is somewhat unknown but cannot be ignored as more and more are being utilized and/or accessioned. Biodegradable plastics, however, are made to break down from environmental exposure; even in a stable museum environment, these plastics will likely be more vulnerable to degradation (Williams 2019).

Figure 3.6 visualizes the development of some common plastics over time; from semi-synthetics to fully synthetics to recyclable to biobased forms.

Sidebar 2.2 3D-Printed Plastics

Meredith Sharps Noyes

3D-printed materials can be found in a variety of applications within cultural heritage institutions. Many artists are using 3D printing as a medium, and fine art and other objects, including architectural models and jewelry, are entering collections (see Sidebar 8.3). In several preservation and access activities such as conservation treatment, storage management, and exhibition, many objects are too delicate to handle or place on display and can benefit from non-invasive, digital approaches aimed at reducing handling, including digital scanning and 3D printing. Examples range from the creation of custom mounts and supports, to replicas of objects, and object fills, integrations, or replacement parts. Across all of these preservation and access applications, 3D-printed plastic has become more common.

3D printing, also referred to as additive manufacturing, describes the fabrication of objects from 3D model data through a layer-by-layer manufacturing process. This definition is broad, and the term encompasses a number of technologies that build objects in different ways. For plastics, common technologies include the following:

- **Fused Deposition Modeling** (FDM), where plastic filaments such as PLA, acetonitrile butadiene styrene (ABS), and PET glycol (PET-G) are extruded through a heated nozzle
- **Stereolithography** (SLA), a process in which a vat of liquid photopolymer is selectively cured
- **Selective Laser Sintering** (SLS), in which powdered plastic particles, most commonly polyamide (nylon), are fused together with heat from a laser

These represent but a few of the many technologies available, and yet they provide a glimpse at the very different methods through which plastic objects can be 3D printed.

Whether caring for 3D-printed objects or using 3D printing as a tool in collection care, the chemical and physical stability of a 3D-printed material should be considered. The technological variety among 3D printing processes and the types of plastics available present a major challenge to institutions. Both the way in which the object is built and the underlying chemistry of the plastic can affect long-term stability of the material.

Chemically, 3D-printed plastics will face the same types of challenges as other plastic objects in collections, but they will also have additional vulnerabilities depending upon the process through which they are made. For example, plastic filaments made for the FDM process often contain additives, including plasticizers, UV stabilizers, flame retardants, and fillers for strength. Photopolymer resins for SLA are often proprietary to the companies that make them and are formulated to have different properties. While plastic powders for SLS are often pure, the heat from the sintering process can cause objects made with this technology to prematurely break down.

Physically, each 3D process will present its own unique vulnerabilities. The layer-by-layer method through which these objects are made

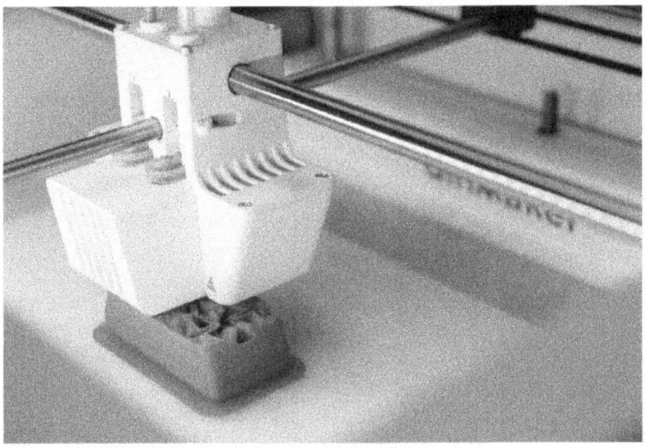

Figure 2.1 An FDM printer creating a plastic object by depositing filament layer-by-layer. Credit: Jaemi Rowe, Rochester Institute of Technology.

often results in internal voids and gaps between the layers, which differ depending on the technology used. These voids can then impact material strength as well as the object's response to external agents such as moisture. Traditionally manufactured or thermoformed structures will not exhibit these same characteristics.

For these reasons, it is extremely useful to know how a 3D-printed object was made (what process or technology), in addition to the type of plastic it is made from. These two pieces of information go hand-in-hand to help predict the preservation vulnerabilities of an object. In collections, identification of 3D-printed plastics remains a challenge so documenting this information at the time of acquisition can assist in understanding how these objects will change over time. In preservation and access, many 3D-printed materials have not been assessed in terms of their safety and compatibility with cultural heritage objects. Understanding the risks of specific 3D printing technologies and the materials used is important in determining whether the technology is appropriate for a given application in a museum setting (such as an object support).

Polymer

Categorizing as natural, semi-synthetic, or synthetic is helpful in understanding groups of plastics but does not fully answer the starting question: What is plastic? Whether fully or partially synthetic, every plastic is a polymer but not every polymer is a plastic. So, what is a polymer? A polymer is a long chain formed by repeating molecular units (monomers) that create macromolecules (polymers). To become a polymer, monomers repeat many, many times (hence the use of poly) and the molecular chain becomes large (hence the use of macro).

The specific monomer used to create the polymer is often reflected in the name of most synthetic plastics, such as polyethylene (ethylene monomer) or polyurethane (urethane monomer).

Cellulose is a polymer but it is not a plastic. However, chemists can use cellulose as part of the formulation for semi-synthetic plastics like cellulose acetate (CA) to create a range of final forms including film, sheets, three-dimensional objects, and fibers. In this process, the ingredients are polymerized. Polymerization is the reaction of monomers to form macromolecules or polymers. The number of monomers that create the macromolecule or polymer is called the degree of polymerization (Williams 2019). Because polymers are large, their molecular weights are large and this size impacts physical properties (van Oosten 2022). Generally, as molecular weight increases, the density, strength, melting point, and boiling point increase as well. Or, in other words, the degree of polymerization affects the properties of that polymer. Polymer chemists work to manipulate these properties to create the desired final plastic.

There are two basic methods for polymerization: addition and condensation. In addition polymerization, monomers are joined without losing simple molecules like water. Examples include PE, PP, and polystyrene (PS). For condensation polymers like cellulose nitrate (CN), CA, and PUR, the linking of the monomers in polymerization produces water or other simple by-products. For collection care, it is helpful to know that addition polymers are not sensitive to moisture while condensation polymers are and, as a result, are vulnerable to high relative humidity and breakdown in a degradation process called hydrolysis (see Chapters 4 and 5) (Williams 2019).

Most synthetic polymers are invented on either side of WWII. Rather than focusing on inventing new plastics, polymer chemists during the last decades of the 20th century and into the 21st century primarily focus on altering about 50 classes of plastics from which they can derive tens of thousands of plastic products (Shashoua 2008).

Plasticity

Beyond every plastic being a polymer, is there a unifying property of all the materials under the broad term plastic? The answer to that stems from the name itself: plasticity. The ability to be relatively easily shaped or molded is the defining characteristic for what makes a plastic a plastic. Whether or not a plastic retains plasticity is another way to categorize plastics into three forms:

- Thermoset
- Thermoplastic
- Elastomer

Tables 2.2–2.4 list plastics in these categories, their polymer abbreviation, and examples of product names. Because names can vary by manufacturer and country, the list of names is not exhaustive.

Table 2.2 Thermoset Examples

Thermoset	*Abbreviation*	*Names*
Casein formaldehyde	CF	Galalith, Lactoid, Erinoid
Epoxy	EP	Araldite, Hxtal
Melamine formaldehyde	MF	Melamine, Formica
Phenol formaldehyde	PF	Bakelite (with or without filler), Catalin (resin, no filler)
Unsaturated polyester	UP	Fiberglass resin, Polylite (*different than PolyLite PLA*)
Urea formaldehyde	UF	Beetle, Beatl, Bandalasta, Linga Longa

Sources: van Oosten (2022); Plastic Identification Tool (n.d.).

Table 2.3 Thermoplastic Examples

Thermoplastic	*Abbreviation*	*Names*
Acrylonitrile butadiene styrene	ABS	Lustran, Cycolac, Novodur
Acrylonitrile styrene acrylate	ASA	Luran S, Dialac
Cellophane	–	–
Cellulose acetate	CA	Lumarith, Celanese, Tenite, Safety Film
Cellulose acetate butyrate	CAB	Cellidor, Tenite Butyrate, Tenite II, Uvex
Cellulose acetate propionate	CAP	Cellidor, Tenite Propionate
Cellulose nitrate	CN	Celluloid, Parkesine, Xylonite, Nitrocellulose, French Ivory, Ivoroid
Ethyl cellulose	EC	ETHOCEL
Polyamide	PA	Nylon, Dartek, Kevlar
Polycarbonate	PC	Lexan, Makrolon
Polycaprolactone	PCL	Polymorph, ShapeLock
Polyethylene (high density)	HDPE	Ethafoam, Tyvek, Volara, Plastazote
Polyethylene (low density)	LDPE	Tupperware, Sterilite (also as PP), Ziploc
Polyethylene terephthalate	PET or PETE	Mylar, Melinex, Dacron, polyester
Polyhydroxyalkanoate	PHA	–
Polyhydroxybutyrate	PHB	Biopol
Polylactic acid	PLA	Polylactide, Plantic, Lactel, PolyLite PLA
Polymethyl methacrylate	PMMA	Plexiglas, Perspex, Lucite, Acrylite
Polyoxymethylene	POM	Acetal, polyacetal, polyformaldehyde, Delrin
Polypropylene	PP	Tupperware, Sterilite (also as LDPE)
Polystyrene	PS	Styrofoam
Polytetrafluoroethylene	PTFE	Teflon, Gore-Tex
Polyvinyl acetate	PVAc	White glues, hot melt adhesives
Polyvinyl chloride (plasticized)	PVC	Tygon
Polyvinyl chloride (unplasticized)	PVC	Sintra
Polyvinylidene chloride	PVDC	Saran Wrap (*PVDC use stops in 2004*)
Styrene acrylonitrile	SAN	Luran, Tyril, Kostil

Sources: van Oosten (2022); Plastic Identification Tool (n.d.).

Table 2.4 Elastomer Examples

Elastomer	*Abbreviation*	*Names*
Balata rubber	–	–
Butadiene rubber	BR	Polybutadiene, Buna
Chloroprene	CR	Polychloroprene, Neoprene
Ethylene propylene diene monomer	EPDM	Vistalon, Dutral
Ethylene vinyl acetate	EVA	Elvax, Ateva
Gutta-percha	–	–
Hard rubber	HR	Vulcanized rubber, Vulcanite, Ebonite
Isobutylene isoprene rubber	IIR	Butyl rubber, Butyl, Esso
Isoprene rubber	IR	Polyisoprene
Natural rubber	NR	Latex
Nitrile butadiene rubber	NBR	Nitrile, Buna-N
Polyurethane (thermoplastic)	PU or PUR	Lycra, Spandex, Elastane
Polyurethane (thermoset)	PU or PUR	*Ester and ether foams*
Polyvinyl chloride (plasticized)	PVC	Tygon
Silicone	SI	Silastic
Styrene butadiene rubber	SBR	Buna-SE
Styrene-ethylene-butylene-styrene	SEBS	Kraton G
Synthetic rubber	SR	*Generic collective term for modern rubbers*

Sources: van Oosten (2022); Plastic Identification Tool (n.d.).

For some plastics, plasticity only pertains to the time of manufacture; once made, the forms are set. Not surprisingly, this category of plastic is called **thermoset** (Table 2.2). In thermosets, a mechanism called cross-linking occurs which connects one polymer chain to another, essentially setting them into place either through heat and pressure, such as in compression molding, or curing through a chemical reaction, as when casting an epoxy resin. Casting can be used on an industrial scale but is also utilized by artists or for at-home-use. If you have ever used a two-part epoxy adhesive to glue something together, you have made a thermoset plastic. When set, thermosets are hard and brittle, no longer retaining the plasticity that they had going into manufacturing. A familiar example of an early thermoset in museum collections is Bakelite (phenol formaldehyde).

For other plastics, the ability to be repeatedly re-shaped remains after production. Or in other words, they are not cross-linked and set into shape. These plastics are aptly categorized as **thermoplastic** and constitute about 80% of plastics (Table 2.3). For museums, this property of retained plasticity means that heat, such as storage in high temperatures, and pressure, even from a low but constant force like gravity, can lead to deformation over time for some thermoplastics. The impact of environment and the importance of choices for storage and exhibit supports are covered in Chapters 5 and 6.

The production options for thermoplastics are more varied than for thermosets and these plastics can be extruded, molded, and foamed (van Oosten 2022). Both thermosets and thermoplastics can be injection molded which is worth noting because this technique enabled mass production of plastics and remains the most common manufacturing process for plastics today (Shashoua 2008). The telltale sign of injection molding is the sprue from the point of injection of the polymer resin into the mold and can often be found on the bottom of the plastic, as seen in Figure 3.4 (van Oosten 2022).

Because they are able to be re-worked under the right conditions, recyclable plastics are all in the thermoplastic family, and when used commercially, the type of plastic gets marked on the plastic itself with a numbered Resin Identification Code (Chapter 3). Though all recyclable plastics are thermoplastics, not all thermoplastics are recyclable and not all recyclable plastics are actually recycled even if disposed of with that intention. In fact, only about 7% of plastics get recycled (Shams, Alam, and Mahbub 2021). Recycled products are often of a lower quality than plastics made from virgin resins and, as a result, have a more limited lifespan and may be more prone to releasing reactive degradation products as they deteriorate (van Oosten 2022; Curran et al 2014). Most plastics can only be recycled into their original form a few times, meaning that that plastic water bottle you put into recycling does not go into an infinite loop of reprocessing. If the bottle remains out of the landfill, it may eventually get downcycled into a final form that is no longer recyclable. Common products from downcycling include PET bottles made into fibers for fleece jackets and high-density polyethylene (HDPE) milk jugs made into outdoor furniture.

Elastomers or "elastic polymers" have the property of natural rubber – meaning that they bounce back to their original shape once a deforming force is removed. Because of this, the terms elastomer and rubber are often used as synonyms. But it is important to realize that there are natural rubbers (balata, gutta-percha, hard rubber, and latex) as well as synthetic rubbers (the remaining elastomers in Table 2.4). Vulcanization, the process in which rubber and sulfur are cross-linked, was key to the development of this category of plastics for both natural and synthetic rubbers because it improved desired properties (van Oosten 2022). The amount of sulfur added during vulcanization affects the final form as well as how it ages. For example, the high sulfur content of hard rubber creates a rigid product but its release can negatively impact nearby materials (Chapter 4).

Elastomers are cross-linked but the amount of cross-linking varies so that when they have more cross-links, they act more like a thermoset, and when they have fewer cross-links, they are more like thermoplastics (van Oosten 2022). Or in other words, elastomers can be made to be rigid, as is the case with gutta-percha and hard rubber, or manipulated to be elastic to the degree desired for the final product, as seen in the myriad of synthetic rubbers in Table 2.4. If flexible, the ability for elastomers to stretch or flex and return to their original shape is reduced over time by the accumulation of stress from use and because of any specific sensitivity to degradation catalyzed by factors such as light, heat, oxygen, and moisture. Picture a once flexible bicycle tire that is now rigid and cracked. The stress from use as well as environmental exposure contributed to its degradation. For museums, knowing that elastomers can lose their elasticity over time should inform storage and display decisions.

Thermoplastics: Semi-Crystalline or Amorphous?

Thermoplastics themselves can be divided further based on the degree to which the polymer chains line up in an organized way or are disorderly. If the polymer chains align, they form crystalline regions and these thermoplastics are considered semi-crystalline. These regions give the plastic strength but because some amorphous areas exist, hence the categorization as semi-crystalline, the plastic has some flexibility and give. In addition to strength, the ordered crystalline regions scatter light so these polymers are usually opaque (van Oosten 2022).

Thermoplastics, whose polymer chains are randomly intermixed rather than having ordered regions, are amorphous. Amorphous thermoplastics are transparent since light can pass through the opened spaces rather than being scattered by crystalline areas. This amorphous composition allows for diffusion so they are more vulnerable to solvents, which you may have experienced if you ever spilled acetone on a CD case made of polystyrene (PS) or attempted to clean a polymethyl methacrylate (PMMA) (more commonly called acrylic or by the tradenames Plexiglas, Perspex, or Lucite) display case with some solvent-based cleaners.

An example of the effect of the proportion of crystalline to amorphous areas in a polymer that you are likely familiar with is the difference between PE bags. Plastic grocery bags, which tend to be opaque or only slightly translucent and strong, are typically made from HDPE which can have up to 80% of the polymer matrix as crystalline which explains its opacity and strength. Dry cleaning bags and zipper-topped storage bags, whether used for food storage or museum collections, are typically made from low-density polyethylene (LDPE). LDPE can have around 65% amorphous areas making them transparent or at least highly translucent and relatively weak (van Oosten 2022). See Chapter 6 for more on PE products used in collection care.

Additives

If plasticity is a defining property of plastic and every plastic is a polymer, do those two factors answer what a plastic is? Not quite... and it is in exploring this question further that the variability of plastic, even in the same plastic category or with the same polymer name, is fully uncovered.

Plastics are more than their polymer base. The additives manufacturers put in affect how the plastic is processed and how the final product looks, behaves, and ages. Common categories of additives and their purpose include the following:

- **Antioxidants** (slows reactions with oxygen)
- **Antistatic agents** (reduce static charge during manufacture and to prevent dust accumulation from static charge build-up)
- **Blowing agents** (foaming process to create cellular structure such as with PUR foams or PE foam planks like Ethafoam)
- **Colorants** (colors all the way through using pigments or soluble dyes)
- **Fillers** (add bulk or reinforcement and may be used to reduce cost)

- **Fire retardants** (helps plastics meet fire safety codes)
- **Heat stabilizers** (slows reactions driven by high temperatures)
- **Lubricants** (processing aids that can help with flow and release from molds)
- **Plasticizers** (helps plastic polymer to flex and have plasticity)
- **Processing aids** (helps in manufacturing. Examples: lubricants, anti-slip agents, surfactants)
- **UV stabilizers** (added to retard the absorption of UV wavelengths of light)
- **Waterproofing** (added to retard water penetration)

Sidebar 2.3 Textile Additives

Kris Cnossen

Human-made fibers often include additives, such as delustering agents, UV stabilizers, fire retardants, waterproofing, and lubricants. Textiles may include coatings or finishings; this is especially true for garments. All of these additives may affect the long-term preservation of the textiles. As composite objects, garments can include plastics as not only human-made fibers but also buttons, zippers, embellishments, and sequins. Many garments have linings made from human-made fibers, even if the front fabric is a high-end material like wool.

It is important to remember that each type of additive has a myriad of options that could have been used and that manufacturers change additives based on price, availability, or new developments. For example, between 1930 and 1960, hundreds of new plasticizers became available (Mulder and Knot 2001). Today, phthalates are the main source for plasticizers (King, Grau-Bové, and Curran 2020). Because phthalates are derived from phthalic acid, they themselves can be a source of acidic issues in storage, as covered in the PVC section of Chapter 4. And, as discussed in Sidebar 2.4, phthalates and other additives bring with them health concerns.

Sidebar 2.4 Potential Health Impacts and Precautions for Handling Plastics

Kerith Koss Schrager

Toxicological and epidemiological studies over the past few decades have raised some alarming concerns about health effects related to exposure to plastics, primarily microplastics and additives which can leach into skin, food, and the environment. Recent focus has been on plasticizers and persistent "forever chemicals" demonstrated to be endocrine disruptors and implicated in the rising incidences of obesity; diabetes; inflammation; and neurological, immunological, reproductive, and fertility issues. Health advocates propose that no dose of this class of chemical is safe because of the body's incredible sensitivity to hormones as compared to the more traditional dose response model in which small amounts of toxic chemicals can be tolerated up to a particular threshold (EPA 2024; Endocrine Society 2024; Loria 2024). Combined with evidence supporting carcinogenic, reproductive, and neurotoxic effects of the thousands of plastic chemicals, these studies raise the question: should collection care workers be concerned about their exposure during the handling of plastics?

Health and safety concerns with the care of degraded plastics have focused on sensory cues such as smell as an indicator of volatile organic compounds (VOCs) or more explicit physical safety concerns (e.g., the explosive nature of degraded cellulose nitrates) (Curran et al 2014; NFPA 2024). While the cracking, sticky, and weeping surfaces observed on plastic objects have long been generally classified as the result of leaching plasticizer, few published studies have chemically identified these compounds on cultural heritage objects or reference them as potential health and safety hazards (Ledoux et al 2023; Tsang 2010).

Additives, including a wide range of chemicals such as plasticizers, flame retardants, UV stabilizers, colorants, and per- and polyfluoroalkyl substances (PFAS), are mostly physically or only loosely chemically bound within the polymer structure. They are therefore more readily leached from the plastics and have been the focus of toxicology and exposure data (Luo et al 2022). However, the safety concerns of the polymers should not be discounted as the chemical degradation of plastics also leads to the production and exposure to micro- and nanoplastics. Due to both their composition and size, these particles potentially pose health risks associated with respiration and skin absorption, leading to the accumulation of these microscopic particles in organs and tissues (Vethaak and Legler 2021).

Exposure limits for chemicals and particles, which help understand the relative toxicity and how to control exposure, either do not yet exist for most plastic additives or may be difficult to apply to the unique exposure scenarios of working with cultural heritage. Occupational exposure limits and regulations, or even toxicological results related to general population exposure, depend on consistent exposure to a particular chemical. However, among collections workers, exposure can range from intimate contact with a single object for a limited amount of time during a conservation treatment to more extended and variable contact for days to years in the case of a collection survey.

Identification of a particular chemical on an object does not imply the level that any given person interacting with that object may absorb. The effects of exposure to multiple chemicals acting in concert, a more realistic exposure scenario, are also unknown. In addition, the concentrations presenting on the surface of a degraded plastic may be higher than what may be found in drinking water or cosmetics for a general population exposure assessment. Historic collections are also likely to house plastics that contain chemicals that have been banned from commercial products for health and environmental reasons, such as polychlorinated biphenyls (PCBs). There are many more factors that should be taken into consideration when determining the appropriate protocols for handling material with unknown or vague health and safety information (for more guidelines, see Koss Schrager et al 2023).

Routes of exposure for plastic handling are primarily skin absorption from touching objects and contaminated surfaces as well as inhalation since some of these additives are volatile or semi-volatile. Plasticizers, such as phthalates and bisphenol A (BPA), and flame retardants, for example, can be absorbed through skin. Accidental ingestion through improper handling and hygiene protocols (lack of handwashing, eating and drinking in work areas, etc.) may also be a concern especially in hectic or high stress situations such as emergency response or working under deadlines.

Engineering controls, which are more universally applicable for most hazards, should be implemented first. This includes handling objects in areas with adequate ventilation and using fume hoods or trunks when possible. Personal protective equipment (PPE) only protects the user and must be selected specifically for the hazard, which may not be known if analytical testing to identify the composition of the plastics and additives has not been conducted.

When selecting the appropriate PPE, such as respirator cartridges and gloves, manufacturer-specific supplied technical data should be consulted. For example, according to 3M, organic vapor cartridges as well as high-efficiency particulate air (HEPA) filters may be necessary for certain phthalates since they may pose chemical as well as particulate hazards (3M, 2022). Glove manufacturers recommend nitrile gloves, the most common type found in cultural heritage collections, as an appropriate choice for certain phthalates, but there are no data or recommendations for some other chemicals that have been identified in plastics. PVC and latex-based gloves should be avoided since they perform less effectively to prevent breakthrough and themselves pose health effects from exposure. Gloves should be changed frequently to prevent permeation as well as cross-contamination of surfaces.

To prevent further incidental contamination, proper isolation techniques should also be implemented including using disposable or easily cleanable work and storage surfaces. Protective, removable clothing should also be worn to prevent contamination of street clothing and workspaces. Consistent cleaning and housekeeping protocols with HEPA vacuums will also prevent the accumulation of contaminated dusts. All materials that may become contaminated may need to be disposed of as hazardous waste according to institutional and municipal regulations.

Protective eyewear should be splash resistant if liquids are present. Impact-resistant clothing, shoes, gloves, and goggles should be used if objects are heavy, sharp, or pose other physical hazards.

Due to the constantly emerging data related to the health implications of exposure to plastic chemicals, Safety Data Sheets (SDSs) may not reflect current toxicological data and workers should pay particular attention to the source, date, and author when selecting the appropriate source of their safety information. Most importantly, when working with or around plastic objects, particularly those exhibiting signs of advanced degradation such as liquid exudates, handlers should proceed using the precautionary principle: in the absence of conclusive evidence that a chemical is safe, measures to prevent exposure should assume that it has the potential to cause adverse health effects.

For plastics in collections it is helpful to pull back and consider what it means if a family of additives are part of the plastic recipe and, perhaps more importantly, what it means if they were intended to serve a function beyond aiding in the manufacturing process and get used up or altered over time. A prime example of this is that if plasticizers migrate out, the plastic may appear like it is weeping and/or become tacky, and as it loses mass, the plastic may shrink or deform (Figures 4.2 and 4.3). Once much of the plasticizer is lost, the plastic will become brittle and may crack or shatter. Sometimes plasticizer loss occurs undetected meaning the plastic can look fine but can actually be vulnerable to shattering since it can no longer flex.

In some cases, an additive can have dual purposes. For example, triphenyl phosphate (TPP) was sometimes used both as a plasticizer and a fire retardant in CA, as evidenced in a research study on a salesman's kit of 49 differently colored CA samples from the Lumarith line of the Celluloid Corporation strung on a metal chain, seen in Figure 2.2 (Tsang et al. 2009). Since the samples are from the same company and fastened together, it can be assumed that the coupons would have experienced the same environment since their creation in the mid-1920s or 1930s and throughout their time in a museum. While many of the coupons are in good condition, some are showing signs of deterioration with three being severely degraded. It is interesting to note that degradation is not clustered and the worst examples are next to coupons that visually appear to be fine, as seen in Figure 2.3. One coupon is extensively weeping, to the point that it has clearly lost mass and looks wet (Figure 10.1) and two are brittle, deformed, and cracked from plasticizer loss (Figure 2.3). When weighed, the severely degraded coupons weighed almost 30% less than those that were still in good condition, proving compositional loss as part of the degradation process. Scientific analysis showed higher concentrations

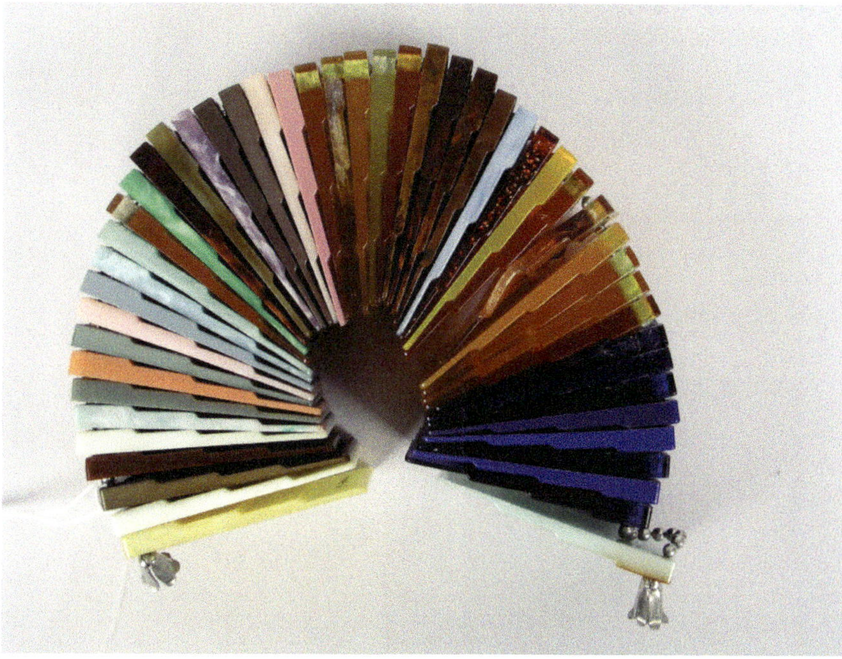

Figure 2.2 Lumarith salesman's kit of 49 CA samples on a metal chain. NMAH 1983.0538.61. Photographed by author, Division of Medicine and Science, National Museum of American History, Smithsonian Institution.

of TPP in the heavily degraded coupons, indicating that in addition to serving as a plasticizer and fire retardant, TPP acted as a degradation catalyst (Tsang et al. 2009). So, not only can the type of additive affect aging but the proportion of it in a polymer recipe can have an impact over time as well. In addition to plasticizer loss, a common degradation phenomenon with CA is the generation of acetic acid in a telltale vinegar smell and evidenced in Figure 2.3 by the corrosion of the metal chain (Chapter 4).

When antioxidants or UV blockers are used up, the result could be discoloration like yellowing. In Figure 2.4, the white plastic has clearly yellowed unevenly. Why? As is often the case, it is helpful to know about the object's history and how the past environment may have affected its current condition. Since this was a white noise sound machine kept in my children's nursery, I can speculate about what happened. The machine was kept on a windowsill meaning that part of it got direct sunlight daily. Over time, the additives meant to retard photo-oxidation, the mechanism in which light and oxygen chemically degrade a material, were exhausted in areas of high-intensity exposure to sunlight, which consists of both the visible light and the more reactive UV light spectrum (Chapter 5). At first the change was subtle and took about one year to appear. You may be thinking, if this was an important

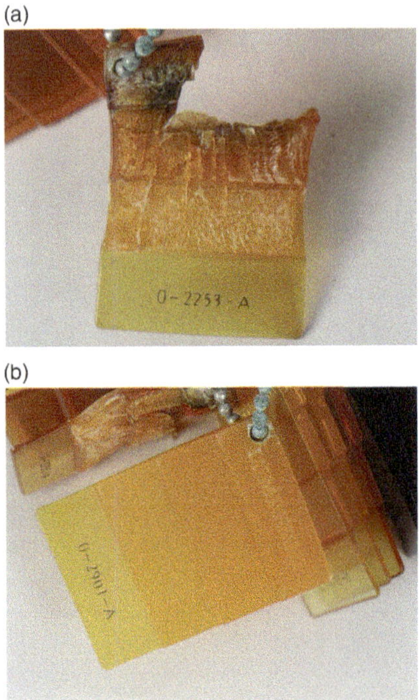

Figure 2.3 Severely deteriorated CA Lumarith coupon (a) alongside intact neighbor (b). Note green corrosion of metal chain. NMAH 1983.0538.61. Photographed by author, Division of Medicine and Science, National Museum of American History, Smithsonian Institution.

museum object, I would put it in dark storage and halt the problem. However, as you can see in Figure 2.4, the darkening continued even after being stored in a dark cabinet, only pulled out once a year as an example for class. It may be surprising that dark storage did not retard discoloration if the catalyst for it is a chemical reaction driven by light and the light source was removed. However, for some degradation mechanisms, their progression is difficult to stop once they get underway even if the environment is improved.

Rather than thinking of additives as secondary components, they have vital influence over not only how a plastic is made, looks, and behaves but how it ages as well, which has obvious implications for museum collections. You may be tempted to think that if a plastic has the same name as another plastic, such as PVC or CN, that all plastics with those names are the same. While they have chemical similarities, namely their polymer, the difference in additives matter and those differences can occur even among the same manufacturer and at similar production times, as seen in the Lumarith CA salesman's kit (Figures 2.2 and 2.3).

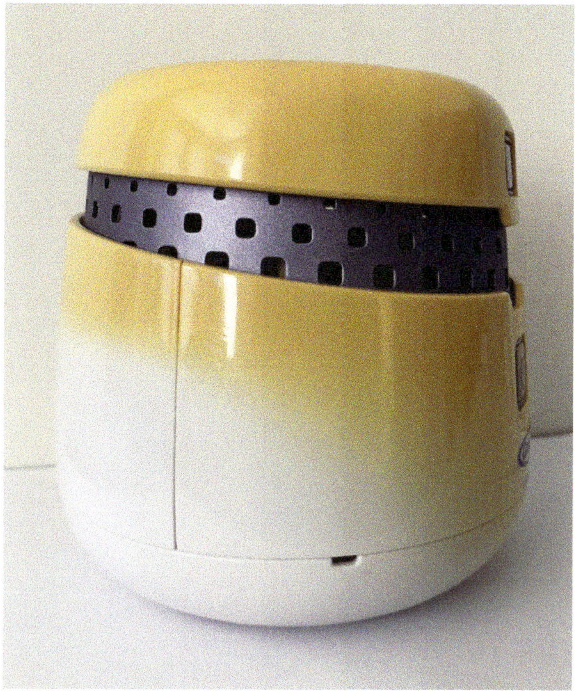

Figure 2.4 Rigid white plastic yellowed from exposure to sunlight and continued darkening after five years stored in a cabinet.

Another example of the impact of additive choices during manufacture could be when different colorant sources are used to produce the same looking final product. Over the anticipated lifespan of the plastic by the consumer, that difference may not be evident but over time with longer than anticipated use or when accessioned into a museum where the desired expectation may be that the plastic looks the same as when it was collected, the difference in formulation may become evident. This scenario played out in the handle of a toy teapot in Figure 2.5. For the first ten years of use, the teapot handle was the same color but eventually a color difference was noticeable. Unlike the white sound machine in Figure 2.4, the teapot was not kept in direct sunlight but rather floated throughout my home with use ebbing and flowing depending on age and interest of my children. Because it is clear that the handle was assembled in two parts, the ready conclusion is that the plastic handle was made in two batches and put together. Again, it looked cohesive when made and a decade run for a toy likely fulfilled the manufacturer's anticipated lifespan of the object. However, the probable different formulations in additives are the most likely culprit to explain the eventual color discrepancy seen in Figure 2.5.

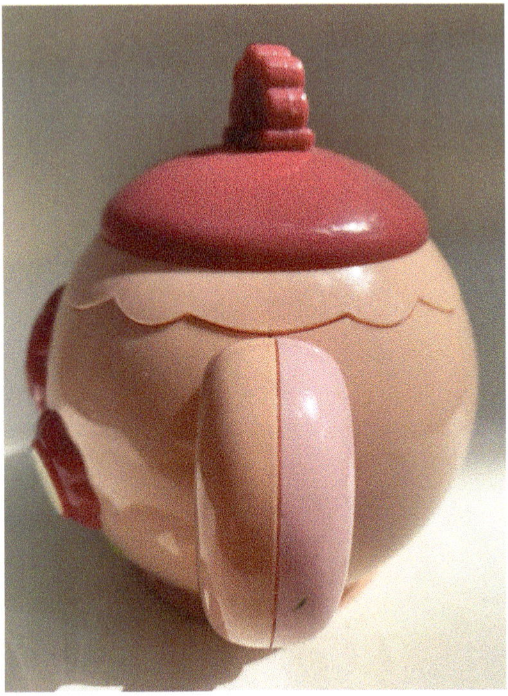

Figure 2.5 Different coloring in the two-part handle of a toy teapot became evident after a decade.

Biobased and biodegradable plastics are seemingly more environmentally friendly because they are not derived from petroleum but remember that they also need additives which can bring with them health and environmental concerns (Sidebar 2.4). That said, some biodegradable plastics are being engineered with the intent of having less of an environmental impact once discarded. An example of this is polyhydroxyalkanoate (PHA) which is a polyester made from microbes digesting fermented sugars or from biogas emitted from landfills or sewage treatment (Baker 2023). Instead of chemical additives, PHA additives are derived from bacteria which act like traditional additives imbuing the desired properties in the final plastic, such as flexibility, but will break down into nontoxic substances. Since both the PHA polymer and the bacteria-based additives are both biobased and biodegradable, it is already being utilized in medicine and agriculture and people are looking to it as a better choice for single use, disposable plastics like food packaging (Baker 2023; Acharjee et al 2023). If these plastics entered a museum collection, how long would they last? Should that be a part of the discussion before accessioning (see Chapter 9)?

Key Points for What Is Plastic?

Every plastic is a polymer and every plastic has plasticity to varying degrees, at least at the start of manufacturing. Whether the plastic retains plasticity helps define it as a thermoset, thermoplastic, or elastomer. Each plastic's characteristics are derived from its formulation: the properties of its polymer base and the additives put in during manufacturing. That formulation, the environment it has been exposed to, and how it has been used all impact how plastics age. So, even though plastics are typically mass produced, each individual plastic object, even if the same type of polymer made by the same manufacturer, is essentially unique. That said, there are general guidelines and considerations to help make decisions for care which will be explored in subsequent chapters. It is important to remember that some plastics can pose hazards to materials in the collection as well as to people. So, it is a good default to always handle plastics with nitrile gloves, unless a different glove type for a specific plastic is deemed safer, making sure to properly dispose of the gloves after use.

References

3M. 2022. 3M Respirator Selection Guide. https://multimedia.3m.com/mws/media/6391 10O/3m-respirator-selection-guide.pdf. Accessed November 2, 2024.

Acharjee, Shiva Aley, Pranjal Bharali, Bhagyudoy Gogoi, Viphrezolie Sorhie, Bendangtula Walling, and Alemtoshi. 2023 "PHA-Based Bioplastic: A Potential Alternative to Address Microplastic Pollution." *Water, Air, and Soil Pollution* 234 (1): 21. https://doi.org/10.1007/s11270-022-06029-2

Arledter, Hanns F. 1964. "Introduction." In *Synthetic Fibers in Papermaking*, edited by Orlando Aloysius Battista, 1–10. New York, NY: Interscience Publishers.

Baker, Aryn. 2023. "The Dirty Secret of Alternative Plastics." *Time*. https://time.com/6339914/plastic-alternatives-pollute/

Chu, Cancy and Petronella Nel 2019. "Characterization and Deterioration of Mineral Papers." *AICCM Bulletin* 40 (1): 37–49. https://doi.org/10.1080/10344233.2019.1672951

Curran, Katherine, Alenka Možir, Mark Underhill, Lorraine T. Gibson, Tom Fearn, and Matija Strlič. 2014. "Cross-Infection Effect of Polymers of Historic and Heritage Significance on the Degradation of a Cellulose Reference Test Material." *Polymer Degradation and Stability* 107: 294–306. https://doi.org/10.1016/j.polymdegradstab.2013.12.019

Dress and Textile Specialists and Plastic Subject Specialists Network. 2022. "Curating Semi-Synthetic and Synthetic Fibres and Fabrics." www.modip.ac.uk/sites/default/files/downloads/Curating Semi and Synthetic Fibres.pdf

King, Rose, Josep Grau-Bové, and Katherine Curran. 2020. "Plasticiser Loss in Heritage Collections: Its Prevalence, Cause, Effect, and Methods for Analysis." *Heritage Science* 8 (123). https://doi.org/10.1186/s40494-020-00466-0

Koss Schrager, Kerith, Anne Kingery-Schwartz, and Julie Sobelman. 2023. "Not a Known Carcinogen: Health and Safety Considerations of New and Innovative Treatments." *Collections* 19 (2): 137–158. https://doi.org/10.1177/15501906231167575

Ledoux, Nicole, Georgina Rayner, Susan Costello, and Angela Chang. 2023. "Preventative Conservation, Treatment and Technical Study of Plasticized Poly(vinyl chloride)

Multiples by Joseph Beuys." *Studies in Conservation* 68 (3): 343–356. https://doi.org/10.1080/00393630.2022.2033520

Loria, Kevin. 2024. How Plastic Can Harm Your Health. Consumer Reports. January 18. www.consumerreports.org/toxic-chemicals-substances/how-plastic-can-harm-your-health-a2854312421. Accessed November 2, 2024.

Lunk, Hans and Karl Stange. 1971. "Synthetic Paper." *Angewandte Chemie Internation Edition* 10 (5): 287–294.

Luo, Hongwei, Chenyang Liu, Dongqin He, Jianqiang Sun, Jun Li, and Xiangliang Pan. 2022. "Effects of Aging on Environmental Behavior of Plastic Additives: Migration, Leaching, and Ecotoxicity." *Science of the Total Environment* 849. https://doi.org/10.1016/j.scitotenv.2022.157951

Mulder, Karel and Marjolijn Knot. 2001. "PVC Plastic: A History of Systems Development and Entrenchment." *Technology in Society* 23 (2): 265–286. https://doi.org/10.1016/S0160-791X(01)00013-6

Mustalish, Rachel. 2007. "Plastics in Paper: History of Synthetic Polymers in Papermaking." Fifth International Conference of the Institute of Paper Conservation and the First International Conference of the Institute of Conservation, *Book and Paper Group*, Edinburgh, 26–29 July 2006. London: The Institute of Conservation, 141–146.

National Fire Protection Association (NFPA). 2024. NFPA 40: Standard for the Storage and Handling of Cellulose Nitrate Film 2025. Quincy, MA. www.nfpa.org/codes-and-standards/nfpa-40-standard-development/40. Accessed November 2, 2024.

Plastic Identification Tool. n.d. "Plastics." Cultural Heritage Agency: Ministry of Education, Culture, and Science. Accessed May 31, 2024. https://plasticidentificationtool.nl/

Quye, Anita. 2014. "Factors Influencing the Stability of Man-Made Fibers: A Retrospective View for Historical Textiles." *Polymer Degradation and Stability* 107: 210–218. https://doi.org/http://dx.doi.org/10.1016/j.polymdegradstab.2014.03.002

Shams, Mehnaz, Iftaykhairul Alam, and Md Shahriar Mahbub. 2021. "Plastic Pollution during COVID-19: Plastic Waste Directives and Its Long-Term Impact on the Environment." *Environmental Advances* 5 (October 2021). https://doi.org/10.1016/j.envadv.2021.100119

Shashoua, Yvonne. 2008. *Conservation of Plastics: Materials Science, Degradation and Preservation.* Oxford: Elsevier/Butterworth-Heinemann.

Tsang, Jia-sun. 2010. "Safe Handling of Plastics in a Museum Environment." *WAAC Newsletter* 32 (2): 16–22.

Tsang, Jia-sun, Odile Madden, Mary Coughlin, Anthony Maiorana, Judy Watson, Nicole C. Little and Robert J. Speakman. 2009. "Degradation of 'Lumarith' Cellulose Acetate: Examination and Chemical Analysis of a Salesman's Sample Kit." *Studies in Conservation* 54 (2): 90–105. https://doi.org/10.1179/sic.2009.54.2.90

United States Environmental Protection Agency (EPA). 2024. Overview of Endocrine Disruption. www.epa.gov/endocrine-disruption/overview-endocrine-disruption. Accessed November 2, 2024.

van Oosten, Thea. 2022. *Properties of Plastics: A Guide for Conservators.* Los Angeles: Getty Conservation Institute.

Vethaak, A. Dick and Juliette Legler. 2021. "Microplastics and Human Health." *Science* 371: 672–674. DOI:10.1126/science.abe5041

Williams, R. Scott. 2019. "Plastic Storage Products." In *Preventive Conservation: Collection Storage*, edited by Lisa Elkin and Christopher A. Norris, 753–799. New York, NY: Society for the Preservation of Natural History Collections.

3 Which Plastic?

Clues to Identification

Even though each plastic object should be thought of as unique, it is helpful to know the type of polymer a plastic is composed of and any information available about its additives because some plastics are prone to severe degradation as they age, pose risks to other materials, and can be vulnerable to specific environmental conditions. In this chapter by covering options that can help identify or at least broadly categorize plastics, collection caretakers can make more informed decisions for storage, display, and loans. Since most collections are cared for in institutions that do not have access to scientific analysis, methods for looking for clues about narrowing down the probable identification through accession records and from the plastic itself are the focus. However, a broad overview of the more frequently utilized options for scientific analysis for plastic identification or classification is included to understand options in case analysis is sought.

Scientific Analysis

When access to scientific analysis is available, there are options for plastics identification or classification to know about since these more conclusive results will help make more fully informed decisions. Some analytical methods require samples of varying sizes and are considered destructive even when the sample is very small because the removed sample is not reattached to the object after testing and, in some cases, the sample is destroyed or so altered that it cannot be reused for other tests. When testing occurs with a small amount of chemical reagent, they are called microchemical spot tests. In other cases, the samples are analyzed with specialized scientific equipment, typically by a conservation scientist. Non-destructive instrumental techniques can be another option and work by accessing the surface of the plastic so do not require a sample.

Scientific Analysis: Spot Tests

If you suspect cellulose nitrate (CN) is in your collection, you want to be wary since it is a problem plastic (Chapter 4). A relatively simple and accurate spot test can be performed to identify CN using a mixture of diphenylamine, water,

DOI: 10.4324/9781003489658-3

and sulfuric acid to create a reagent (so this test obviously can only be performed in a laboratory setting with all the required safety protocols such as a fume hood and necessary personal protective equipment (PPE), namely splash goggles, a lab coat, and nitrile gloves). The reagent is then either directly applied to a sample of CN or by obtaining a dusting of it by rubbing the frosted section of a microscope slide across the surface. One drop of the diphenylamine reagent is then applied to the sample. If nitrates are present, the sample will instantly turn dark blue, as seen in Figure 3.1. By knowing which objects are CN, you can make decisions for monitoring, storage, and display (Chapters 4–7).

An interesting side note is that while diphenylamine spot tests are very reliable in identifying CN, there is the possibility of a false positive in other types of plastics because of exposure to deteriorating CN. Like other materials, plastics interact with and are impacted by their environment. For some, this will result in yellowing, as seen in Figures 2.4 and 6.1. For others, it can mean taking on reactive products released as other nearby materials deteriorate. For example, polyethylene (PE) is a plastic commonly used to house collections (Chapter 6) and is adsorbent (Chapter 4). When a drop of diphenylamine reagent was applied to a section of a PE bag that was ill advisedly used to store a degrading CN object, a positive reaction occurred (Figure 3.1). Degradation products were adsorbed by the PE enclosure to the extent that it now contained enough nitrates to get a positive reaction with the diphenylamine test. This does not mean that the reagent was off or that the diphenylamine test can also be used to identify PE. Rather, it serves as a reminder that plastics can be adsorbent and permeable (Chapters 5 and 6) and that it is vital to utilize all the information at your disposal to make a holistically informed identification rather than just relying on one factor.

Another option for CN identification is MQuant Nitrate Test papers which is a semiquantitative test that uses a color reagent on a test stick to determine if nitrate ions are present and, therefore, whether an object is CN. The advantage of this test is that no sample is required and the test sticks are readily available. However, if

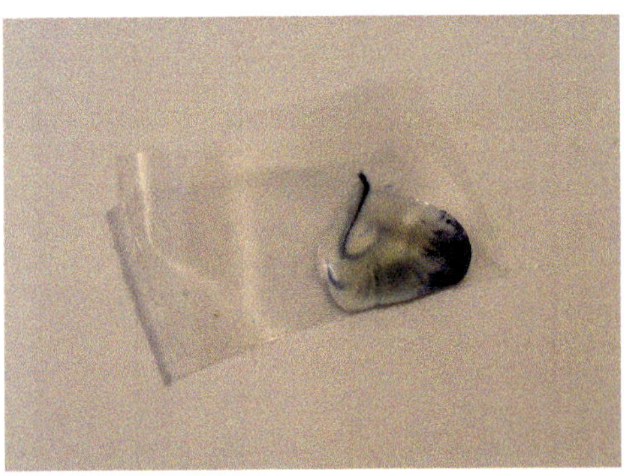

Figure 3.1 Positive reaction of diphenylamine on polyethylene bag used to store degrading cellulose nitrate.

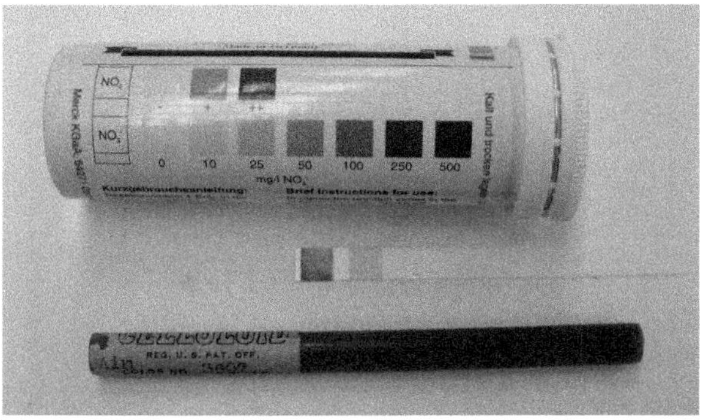

Figure 3.2 Positive reaction of Nitrate Test paper with plum colored "Celluloid" cellulose nitrate rod from 1941. NMAH 1983.0538.91. Photographed by author, Division of Medicine and Science, National Museum of American History, Smithsonian Institution.

left in place too long, the dye on the strip can stain the object's surface so testing should be done in an inconspicuous location. For some objects, the risk of dye transfer may be too great to warrant use.

Nitrate Test papers procedures:

- Store papers in refrigerator until use
- Put two drops of deionized water on surface of plasic
- Apply test strip to wetted surface and let sit for 20–30 seconds
- Remove strip
- Let strip develop for about 40 seconds
- Compare strip color to chart on container
- Positive reactions vary from light pink to dark purple (color chart on container)

Nitrate Test papers factors to consider:

- Deteriorated CN objects react faster and stronger – turning dark purple
- Imitation ivory CN have lighter reactions, possibly because they are typically more stable (see Chapter 4)
- Do NOT test on crizzled or cracked areas – water will wick in!
- May take up to a minute to see full color change
- Can write object identifier on test strip to help keep track during testing

Scientific Analysis: Burn Tests

Some older references will describe the utilization of burn tests, sometimes called pyrolysis, to identify plastics. These methods essentially involve burning a very small plastic sample and noting the color, shape, and duration of the flame as well as the color and odor of the produced smoke or how the sample reacted: did it melt, drip, continue to burn, or burn out? Because of the obvious health concerns with burning plastics, even in small quantities, and because these tests are subjective and destructive, the field has moved away from burn tests for plastics identification and is no longer recommended.

One exception persists, however: the Beilstein test. In this test, plastic is melted onto a heated copper wire that is then placed in a flame. A vivid green or bluish green flame denotes the presence of a halogen which is extrapolated to indicate chlorine which is used to denote that the plastic is polyvinyl chloride (PVC). Since PVC is a problem plastic (Chapter 4), knowing if it is in collections as an accessioned object or being used in storage, such as for photo sleeves, is important to know. For those that have access to x-ray fluorescence (XRF) for analysis, as described below, identification of the element chlorine can be used to identify PVC and does not require destructive sampling or potential exposure to health hazards from burning plastic.

Scientific Analysis: Instrumental Techniques

Instrumental analysis can provide fuller information about a plastic by identifying its polymer and/or additives. Depending on the information desired, using more than one technique may be necessary. Generally, the most frequently utilized or potentially insightful instrumental techniques for plastics are as follows:

- Fourier transform infrared (FTIR) spectroscopy
- Raman spectroscopy
- Pyrolysis-gas chromatography/mass spectrometry (Py-GC/MS)
- Headspace-solid-phase microextraction-gas chromatography/mass spectrometry (HS-SPME-GC/MS)
- X-ray fluorescence (XRF)
- Near-infrared reflectance (NIR) spectroscopy

Fourier transform infrared (FTIR) spectroscopy is the primary analytical technique utilized for plastics. FTIR essentially works by exposing a material to the infrared (IR) light spectrum. Some molecules will absorb the IR light, while in others, the IR light will pass through. The resulting absorption or transmission of the IR light gets plotted into a spectrum which creates a "fingerprint" of the material which then gets compared to a spectra database of known materials. For plastics, the two FTIR techniques that are primarily used are Attenuated Total Reflectance-FTIR (ATR-FTIR) and Specular Reflection-FTIR. The focus for plastics in FTIR analysis is typically to identify the polymer and additives. Matches between spectra can help identify these components or give insight into their degradation, such as if the polymer itself is deteriorated, as seen in Chapter 7 (Figures 7.6 and 7.7).

FTIR analysis can occur either by testing a very small sample in a lab or a portable FTIR can be brought to the object itself to analyze the surface. When testing on the object, slight contact pressure is needed in order to get as accurate a reading as possible. Because contact is needed, an accessible flat area that can withstand some pressure is required, which can be limiting. When a sample is taken, typically the material is flattened between two diamond surfaces and examined under a microscope attachment to best line up areas of interest or to take multiple readings (Richardson and Rivenc 2012). In both methods, results are rapidly obtained.

FTIR can provide information about the polymer itself, changes to the polymer as it degrades, additives, and the products that may be depositing at the surface over time, such as plasticizers leaching out. Though FTIR is considered the standard for plastics analysis, it does have limitations. Difficulty flattening the sample or making adequate surface contact on an object, low quantities of additives present or interferences from them, as well as issues caused by overlapping spectra in complex plastics, such as laminates, can affect results and leave an incomplete picture of the actual plastic composition. Additionally, the results can only be as good as the reference database and interpretation of those results is affected by the experience of the person interpreting the findings (van Oosten and Learner 2012).

Raman spectroscopy is available in both stationary and portable models and results are obtained very quickly. Raman spectroscopy uses a high-intensity laser; much of this light will get absorbed but molecules or groups of molecules, such as in a polymer, will scatter some of the light at different wavelengths and at varying intensity which get plotted into a unique spectrum. The intensity of the peaks in the produced spectrum is proportional to concentration which can give a fuller understanding of the composition of the material. Testing does not require contact and this method is non-destructive; however, dark samples may darken or burn from the laser (Shashoua 2008). As with FTIR, the results are only as good as the spectra database referenced and the person interpreting results.

Pyrolysis-gas chromatography/mass spectrometry (Py-GC/MS) is a destructive technique that can provide a fuller understanding by identifying individual components, such as additives, of a complex sample.

Typically, in the gas chromatography–mass spectrometry (GC/MS) part of this testing, the sample is dissolved into a solution that is injected into the heated gas chromatography (GC) portion of the instrument where the sample is vaporized and mixed with a pressurized inert gas like helium that pushes the sample through a column in which the different components separate out. The properties of each molecule determine how long it takes to move through the column. The molecules then enter a mass spectrometer (MS) which essentially measures the ions produced and the data collected are graphed.

For most plastics, an additional step of pyrolysis (Py) is added to the GC/MS process because synthetic polymers are difficult to put into solution. So, instead, high temperature (pyrolysis) is applied to break the plastic sample into a cloud of smaller molecules that then pass through the GC/MS, revealing more details about its chemical fingerprint, thus becoming Py-GC/MS (Schilling et al 2012). Because Py-GC/MS provides so much data about the sample, its analysis can take more time and requires more detailed interpretation but with that comes more information about additives and degree of deterioration of the polymer (Schilling et al 2012). As with other instrumental techniques, databases of the mass spectra of compounds are utilized for identification.

Headspace Solid-Phase Microextraction Gas Chromatography/Mass-Spectrometry (HS-SPME-GC/MS) uses a coated fiber placed in the "headspace" (HS) around the material being tested in order to collect samples of emitted products such as volatile organic compounds (VOCs). After collection, the Solid-Phase Microextraction (SPME) fiber is inserted into the GC/MS in which the sample is separated by the GC then put through the MS portion of the instrument for characterization (Lattuati-Derieux et al. 2013). Benefits of HS-SPME-GC/MS analysis are that no sample is needed, it evaluates deterioration markers which can alert caretakers to condition issues before degradation is evident, and it can identify emitted products that can affect other nearby materials (King, Grau-Bové, and Curran 2020). This method is completely non-invasive since it is testing the volatiles around the plastic rather than the plastic itself.

X-ray florescence (XRF) is an important analytical method to be aware of not because it has a particular strength for plastics, in fact its use with plastics is limited, but it is a more widely available instrument that is sometimes owned by museums that do not have conservation scientists on staff. The appeal of XRF is that it is non-destructive with no sampling required. For museums without scientific laboratories, the portable handheld units are the most utilized because they can be used to analyze a range of objects in situ, are easy to store, and can be operated by conservators or other trained collection staff.

XRF works by emitting an X-ray that then excites electrons in the atoms of the material being analyzed. When the excited electrons fall back into their original position, they emit fluorescent X-rays that are unique to each element and, therefore, can be used to identify the elements present in a material. The intensity of the graphed peaks essentially corresponds to concentration of those elements so the stronger the peak, the greater the concentration.

The elements able to be detected depend on the sensitivity of the instrument with more modern XRF units being able to detect lighter elements. For example, in the early 2000s most portable units used by museums could not detect chlorine but now detection is possible, meaning that XRF can be utilized to look for chlorine to identify PVC. Besides identifying PVC, use of XRF for plastics is basically limited to identifying inorganic additives such as fillers like calcium carbonate or pigments like zinc oxide.

Near-infrared reflectance (NIR) spectroscopy is a valuable technique for the identification and characterization of plastics. By analyzing the absorption and reflectance of light in the near-infrared (NIR) regions of the electromagnetic spectrum (780–2500 nm), this method detects molecular vibrations characteristic of specific polymer types. It is widely employed for sorting waste plastics in recycling due to its efficiency and non-destructive nature (Ishikawa, Hikima, and Ozaki 2021). However, the technique has limitations. Distinguishing plastics with similar chemical structures, such as blends or co-polymers, presents a challenge due to overlapping spectral signatures, often requiring advanced statistical techniques for confident identification (Šuštar et al 2014). Additionally, analysis is significantly hindered in plastics containing carbon black-based fillers, which absorb strongly across the NIR range impeding detection.

While NIR spectroscopy has limitations and can lack the specificity of certain other analytical methods, it can be particularly effective in museums because the equipment is portable and can be coupled with a fiber optic probe for analysis of objects in situ (Cucci, Bigazzi, and Picollo 2013). For example, during planning for the reinstallation of the Contemporary Galleries at the Detroit Institute of Arts (DIA), vis-NIR fiber optics reflectance spectroscopy was used to successfully characterize most plastics in collection storage by comparison to a reference database. The portability and rapid analysis enabled the DIA Conservation Department to identify materials and make treatment and targeted display recommendations with minimal disruption to workflow (Christina Bisulca, Andrew W. Mellon Conservation Scientist, DIA and Liz Homberger, Objects Conservator, DIA. Personal communication. January 21, 2025).

Non-Analytical Identification

For the majority of institutions that care for cultural heritage, access to scientific analysis is limited or nonexistent. So, what options are available to at least categorize if not fully identify a plastic? The first place to start is the most obvious: examine the plastic itself!

Non-Analytical Identification: Resin Identification Code

Modern plastic markings that we are all familiar with are those used to sort recycling. Taking the form of a pyramid with numbers from 1 to 7 within it, often with the polymer abbreviation below it, is the Resin Identification Code (RIC). The RIC was created in 1988 by the Society of the Plastics Industry as a way to identify the polymer resin used to make a plastic in order to aid recycling. Since 2008, the RIC has been administrated by ASTM International, formerly known as American Society for Testing and Materials (Kelechava 2019).

The RIC divides plastics into seven numbered categories:

1 Polyethylene terephthalate (PET or PETE)
2 High-density polyethylene (HDPE)
3 Polyvinyl chloride (PVC)
4 Low-density polyethylene (LDPE)
5 Polypropylene (PP)

6 Polystyrene (PS)
7 Other

For use in collection care, it is important to remember that these codes are only going to be found on commercially produced products after 1988. The code only gives insight into the polymer and does not give any indication of the additives used to make the plastic product. That said, knowing the polymer can be helpful in making storage and exhibition decisions (Chapters 5 and 6). Since the seventh category is a smattering of "other" plastics, this labeling is only helpful in knowing what a number 7 plastic is NOT, as in it is not a plastic in the other six categories. It may be helpful to know that the most produced plastics today are PE (HDPE and LDPE), PP, PVC, and PET, in that order (Plastic Identification Tool n.d.). In terms of recycling, PET and HDPE are the most frequently recycled plastics.

Non-Analytical Identification: Tradenames and Patents

Some plastics may have trademarks, tradenames, product line names, or other clues marked on the plastic, as was the case with the Lumarith samples discussed

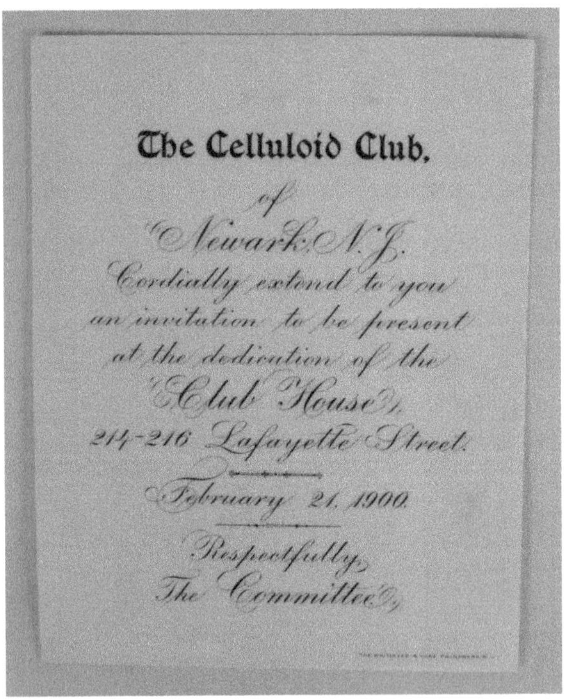

The Celluloid Club,
of
Newark, N.J.
Cordially extend to you
an invitation to be present
at the dedication of the
Club House,
214-216 Lafayette Street.
February 21, 1900.
Respectfully,
The Committee.

Figure 3.3 The Celluloid Club of Newark, NJ invitation, 1900. NMAH CH.334327. Photographed by author, Division of Medicine and Science, National Museum of American History, Smithsonian Institution.

in Chapter 2 and seen in Figures 2.3 and 10.1. The Lumarith example highlights the importance of researching the product type itself and not making assumptions based on the company name. It may seem reasonable to assume that the Celluloid Corporation would only make CN since Celluloid is a tradename for that type of plastic but the company's Lumarith product line is composed of cellulose acetate (CA) and was produced and advertised in a targeted response to combustion concerns with CN (Chapter 4).

Another example of looking for identification clues is in Figure 3.3 in which the Celluloid Club of Newark, New Jersey, is inviting members to the dedication of their clubhouse in 1900. It seems very reasonable to assume that the Celluloid Club would use CN since Celluloid is a tradename for that semi-synthetic plastic and Newark was a center of Celluloid production. However, as seen with the Lumarith example above, the Celluloid Corporation itself did not always limit itself to CN. So, what additional clues are available to aid in identification? The fact that the rigid sheet is meant to imitate ivory (more in Chapter 4) and the date of printing is 1900 (see timeline discussion later in the chapter and Figure 3.6) all point to this object being CN. If possible, you could confirm this identification with the diphenylamine spot test or Nitrate Test papers as outlined in the "Scientific Analysis: Spot Tests" section in this chapter or with instrumental analysis, if available.

Sidebar 3.1 Dating Tupperware

Figure 3.4 shows the bottom of a Tupperware cup marked with "Tupper Corporation Farnumsville, Mass. U.S.A. 122-2." This readily indicates that the translucent flexible cup is a piece of Tupperware and the central sprue mark indicates injection molding, a process discussed in Chapter 2. Tupperware was originally made of PE and the markings narrowed down the date of manufacture from between 1942, when Earl Tupper opened his first Tupper Plastics Factory in Farnumsville, Massachusetts, and 1947, when Tupper was awarded his first patent (NMAH n.d.; Eschner 2018). The number 122-2 on the cup is not a patent number but rather Tupperware's two-part model number (Tupperware 2024). The lack of a patent number, which is marked in subsequent Tupperware products either with the patent number, trademark or patent symbol, or the words "Tupper Seal" (sometimes with the profile of a seal to indicate the patented seal technology), combined with the stamped name of the first Tupperware factory, as well as the lack of the name of a product line like "Welcome Ware," "Wonderlier," or "Servalier," all point to this being an early example of Tupperware, possibly dating to WWII (NMAH n.d.). So, in this example, dating and identification had as much to do with what is there as what is not.

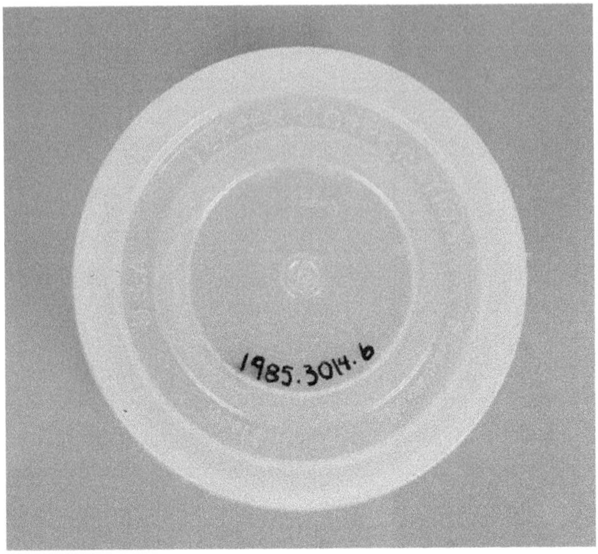

Figure 3.4 Bottom of Tupperware cup marked "Tupper Corporation Farnumsville, Mass. U.S.A. 122-2". NMAH 1985.3014.06. Photographed by author, Division of Home and Community Life, National Museum of American History, Smithsonian Institution.

If markings from the manufacture exist but do not point to an obvious or likely plastic, you can do some research to look for potential matches. If a patent number is indicated on the plastic, searching the patent registry of the likely country of manufacture is a great place to start.

The Plastics Historical Society has an alphabetized listing of trade names with their associated polymer type and manufacturer. There are also a few trademark symbols sketched on their website. There are two important limiting factors to realize for this resource though: the content is primarily focused on plastics before 1933 so very few fully synthetic plastics are included (Figure 3.6) and the listed manufacturers are concentrated in Europe and the United States so plastics made in other regions of the world are not included (Plastics Historical Society n.d.).

Non-Analytical Identification: Online Tools

The Museum of Design in Plastics (MoDiP) at the Arts University Bournemouth in the United Kingdom created A Curator's Guide to Plastics website in 2007 "intended for those with little or no prior knowledge of plastics and as a starting point for further exploration" (MoDiP 2007a). This user-friendly, accessible approach contains a lot of helpful information for collection caretakers, including for researching tradenames or getting information on a specific plastic from the "Plastic Materials" page. The links from that page lead to data sheets that have representative images of the plastic, tradenames, characteristics, common uses, and information on degradation. If you are not sure where to start, there is an "Identification route map" which is essentially a yes/no flowchart to help narrow down possible plastic types for identification based on observable characteristics and any known dating information. The questions can be answered online or a flowchart can be printed. For those with semi-synthetic and/or synthetic textiles to care for, the site also has a section titled "A Curator's Guide to Synthetic Garments" that developed from a project from 2020 to 2022 with the goal to increase understanding of synthetic textiles among collection stewards.

Another helpful online resource is the Plastic Identification Tool which has an immense amount of accessible and practical information to help guide the care of plastics in collections. It developed from the *Project Plastics* that took place in Europe between 2017 and 2019. The site has a section on its website called "List of Names" which provides tradenames or brand names listed in alphabetical order along with the polymer type that constitutes products under those names. Clicking on the polymer name will take you to a data sheet on that plastic (outlined below). Or conversely, you can navigate by the type of plastic, if known, and then the linked data sheet with common tradenames or brand names will appear.

If you do not have a name to start with, the Plastic Identification Tool online resource is still a viable tool you can use by first deciding which of four broad categories best describes the plastic in question:

- Foam
- Film
- Elastomer
- Rigid

The site further categorizes rigid three-dimensional objects as sheet, industrial product, or made by artist with this last category having the most unknown variables. Some polymers that can take rigid final forms can also be utilized as films. This site defines a film as "…thinner than 1 mm and you can roll it over a tube. There is no preformed shape (vacuum drawn or mold-made)" (Plastic Identification Tool n.d.).

If your entry point to using the Plastic Identification Tool is choosing one of the four broad categories, you are then taken to a questionnaire to input answers about appearance, possible date of manufacture, markings, signs of manufacture, deterioration, etc. If you can touch the object or take a sample for testing, additional questions will appear. The questions are dynamic so as you answer, other questions may populate. The ranking score of possible plastic candidates for the object is listed on the side and will rearrange so that the likeliest plastics move up the list as questions are answered. You can then click on the plastic types in the ranked list and go to the plastic data sheet. Remember, though this is a powerful tool, it is only as good as the data inputted and the contextual understanding of the object in question by the user. Or, in other words, misidentifications are quite possible! If plastic identification from using the tool is recorded in collection records, be sure to note the source and the degree of confidence in the identification.

The plastic data sheet content in the Plastic Identification Tool is divided into six sections:

- General Information
- Other Names
- Am I Dealing With …
- Properties
- Degradation
- Preventive Conservation

The "General Information" section contains a description or general overview, the history of when the plastic was invented and when it went into commercial production, followed by information on its manufacture, application, and appearance. To the left of this content are a few representative images and the chemical structure of the polymer. To the right is a list of "Other Names" which are the aforementioned tradenames or brand names for that plastic type. The content in "Am I Dealing With…" is important because the same plastic polymer can be manufactured into different forms: foam, film, elastomer, or rigid as previously discussed.

Next, "Properties" includes two sections: Material Properties and Identification Properties. In Material Properties, chemical information like melting point, when a solid turns into a liquid, is provided. For amorphous plastics (see Thermoplastics section in Chapter 2), the melting point is a range, while semi-crystalline plastics have an exact melting point (van Oosten 2022). In this section, glass transition temperature (Tg) is also given. Tg is the temperature at which the plastic shifts from hard and rigid to flexible and soft. In practical terms, plastics with low Tg are vulnerable to changes if stored at high temperatures, such as in a hot attic. Some archaeological collections may have experienced the effect of high temperature on plastic adhesives with low Tg used to reassemble pottery fragments. If done in a hot field environment or stored at high temperatures, the adhesive can become soft and rubbery causing the pottery to slump as the joins fail.

The "Properties" section also has information on identification. This includes a description of cell structure (if a foam); how the plastic may smell; what it feels like when touched (such as if it feels cold); sound when struck (for example, some plastics sound "tinny" when tapped); ultraviolet (UV)-radiation, indicating how a clear plastic of that type fluoresces under UV light; and how a transparent plastic, either in rigid three-dimensional form or as a film, looks when viewed under polarizing filters, as discussed in Sidebar 3.2 and Tables 3.1 and 3.2.

Sidebar 3.2 Polarizing Filters and Birefringence

Mary Coughlin and Kris Cnossen

To understand more fully the information gleaned from the section on polarizing filters in the Plastic Identification Tool, it is helpful to know a few points. The colors or patterns that are produced when a transparent material is viewed through polarized filters are called birefringence. If you have ever worn polarized sunglasses in a car with tinted windows and noticed colored dot patterns on the windows when the sun hits at a certain angle, you have experienced this effect. You do not need specialized equipment to test for birefringence; all you need is a pair of polarizing sunglasses, the transparent plastic you are examining, and an even, solid light source like a lightbox or a white screen on your computer, such as a blank document. Place the object against the solid light backdrop and put the polarized lens in front of the object. You will instantly know if colors are present or not (Figure 3.5). It can be more difficult to decide if gray/brown/white birefringent patterns are present but this is a characteristic of some plastics (Table 3.1). If you see patterns, rotate the lens; the colors or

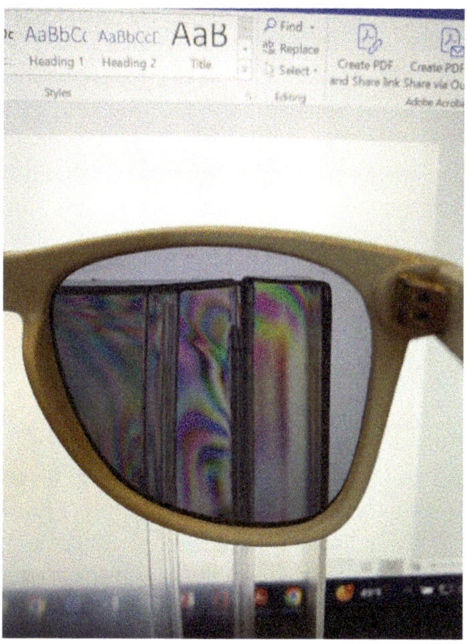

Figure 3.5 Polypropylene box (marked with RIC #5, PP) propped against a blank Word document with polarized sunglasses in front producing bright birefringence colors.

Table 3.1 Birefringence of Transparent Plastics by Polarizing Filters

Plastic	*Bright Prismatic Colors*	*Gray/Brown/White*
Casein formaldehyde (CF)	x	
Cellophane (CE)	x	
Polybutylene adipate terephthalate (PBAT)		x
Polycarbonate (PC)	x	
Polyethylene (PE) (LDPE and HDPE)		x
Polyethylene terephthalate (PET)	x	
Polypropylene (PP)	x	
Polyvinyl chloride (PVC) FILM *plasticized and non-plasticized*		x

Source: Plastic Identification Tool (n.d.).

patterns will shift and the background light will extinguish. Note that this test is best done on clear plastics but birefringence on transparent colored plastics can also be detected.

The use of polarized filters for examination of transparent plastics is helpful not only for identification, or at least narrowing down the possible plastic types, it can also help indicate points of stress with the darker the color, the more stressed the plastic is in that area (van Oosten 2022). Points of stress may result from use, like plastic hinges, from environmental factors, or from production, such as the point of injection into a mold or pressure from extrusion.

If available, polarized light microscopy (PLM) can be used to aid in the identification of human-made fibers which also may have telltale birefringence. PLM utilizes filters similar to those used in polarized sunglasses. In order to see the birefringent colors of a human-made fiber, the crossed polar filters need to be active, which can be confirmed by the black field of vision. When the stage of the microscope is rotated, the fiber will become dark in certain positions and lighter in others, the lightest position being the 45° angle. Using the 45° angle to determine the birefringence colors can help narrow down the human-made fiber type. The use of refractive index liquids to observe the Becke Line of the fiber's surface will likely be needed to further narrow down the identification of a human-made fiber (Reffner, Kammrath, and Kaplan 2020; Wheeler and Wilson 2008). Examples of fibers under PLM are found in the "Fibre Identification" section of the "Curator's Guide to Synthetic Garments" on the MoDiP's website (MoDiP 2022). If you are not certain of the exact type of fiber but you know it is human-made, either semi-synthetic or synthetic, that is still a helpful classification to include in collection records.

Table 3.2 Birefringence of Human-Made Fibers with Polarized Light Microscopy

Fiber Type	*Birefringence*				
	Dark (no birefringence, isotropic)	*Whites, grays, dull yellows (low biref-ringence)*	*Muted colors of mainly gold with touches of orange and blue (medium biref-ringence)*	*Very hot colors like bright blue, magenta, and hot yellow (high biref-ringence)*	*Faint pastels with pink and green banding (very high biref-ringence)*
Rayon			x		
Acetate		x			
Triacetate	x				
Nylon				x	
Polyester					x
Polyurethane					
Acrylic		x			
Azlon		x			

Credit: Kris Cnossen.

The "Degradation" section of the Plastic Identification Tool essentially covers a plastic's vulnerability to the environmental factors of light (listed as photo-oxidation), humidity (indicated as hydrolysis), or if the produced degradation products accelerate deterioration of the plastic itself (called auto-catalysis). Next, whether or not that plastic is considered a problem plastic is noted under "Details." The noticeable results of deterioration are listed as "Symptoms" and include indicators such as yellowing, warping, formation of crystals, and odor. Next, "Susceptibility," or that plastic's vulnerability to the environmental risk factors of UV-radiation (light in the UV spectrum), light (light in the visible spectrum), oxygen/ozone, temperature, and relative humidity (RH) are ranked as low, medium, or high. Under these rankings there are sometimes narrative descriptions elaborating the environmental sensitivities.

The last section on the Plastic Identification Tool data sheets is "Preventive Conservation" which is of paramount importance to collection caretakers because it provides guidance about the environment to keep the type of plastic. It includes factors that may be controllable like light levels, temperature, and RH ranges. For plastics vulnerable to oxygen or ozone, keeping them in environments with reduced levels is often difficult to achieve and maintain but flagging that as ideal is good to know and may be something that could be worked toward, at least for some items. For some plastics there are additional preventive notes that provide fuller context.

Non-Analytical Identification: Collection Records and Timelines

Institutional records like accession files can aid in narrowing down the possible type of plastic. This information, of course, has to be considered in relationship to the object's appearance and function when in use. For example, if you have a clear film from 1920, you could narrow down your possible plastics to CN, CA , and cellophane since each can be a clear film and were available at that time (Figure 3.6). You would never identify the film in question as Bakelite (phenol formaldehyde) even though it was also available at that time since Bakelite products are dark and rigid. To get further toward narrowing down the plastic film in this example, what was it used for? Is it a film in the sense of a photograph or movie reel or a film used as a wrapping? If the film has images on it, then you can rule out cellophane. For film and photographs, information on notch codes and edge printing, such as from Image Permanence Institute's FilmCare.Org, could be useful aids in determining the base plastic (2024). Otherwise, resources like the Plastic Identification Tool described above can assist in figuring out the likely plastic type. Or if it is still uncertain, you can compare entries for the most likely plastics and determine which environment could best serve the object regardless of knowing its exact polymer.

Names of manufacturers that may be in the accession files, even if not marked on the plastic itself, can be researched as described in the previous sections to figure out what kinds of plastics they made and when, again narrowing down possible plastic types. Referencing timelines can be a great resource to aid this quest because if you know the date of manufacture or even the year of accession into the collection, you can eliminate plastics that came after that date, thus narrowing down your search.

The MoDiP has an annotated timeline on their website that highlights key plastic developments from 1712 to 2007. The plastics listed are hyperlinks that go to pages containing the content described in the "Tradenames and Patents" section earlier in this chapter. The timeline includes various uses of the same plastic type. For example, on the vertical timeline you can find that acrylic (polymethyl methacrylate) was first used in the canopies of Spitfire planes in 1936, then in 1938 the same polymer was used for the first plastic contact lenses, and in 1947 acrylic paint became available. The use of acrylic for plane cockpits, contact lenses, and paints well encapsulates the variety of forms chemists can create from the same polymer. If it is more helpful to search by swaths of time, the site has a section called "When was it made" in which available materials and manufacturing process are listed under blocks of time, like 1925–1940, with hyperlinks to additional information.

When considering possible plastics, it is important to note that there can be a discrepancy, sometimes by several years or even decades, between when a plastic was invented and when it went into commercial production. An example of this is CA which was discovered in 1865 but not commercially produced until 1910, though some products started to be made around 1904 (Plastic Identification Tool n.d.). The timeline in Figure 3.6 marks dates of commercial production, not invention. The date of manufacture can vary by country so these dates are not absolute and sometimes when the development was protracted and discoveries overlapping, finding a single recognized date of manufacture can be difficult, as is the case

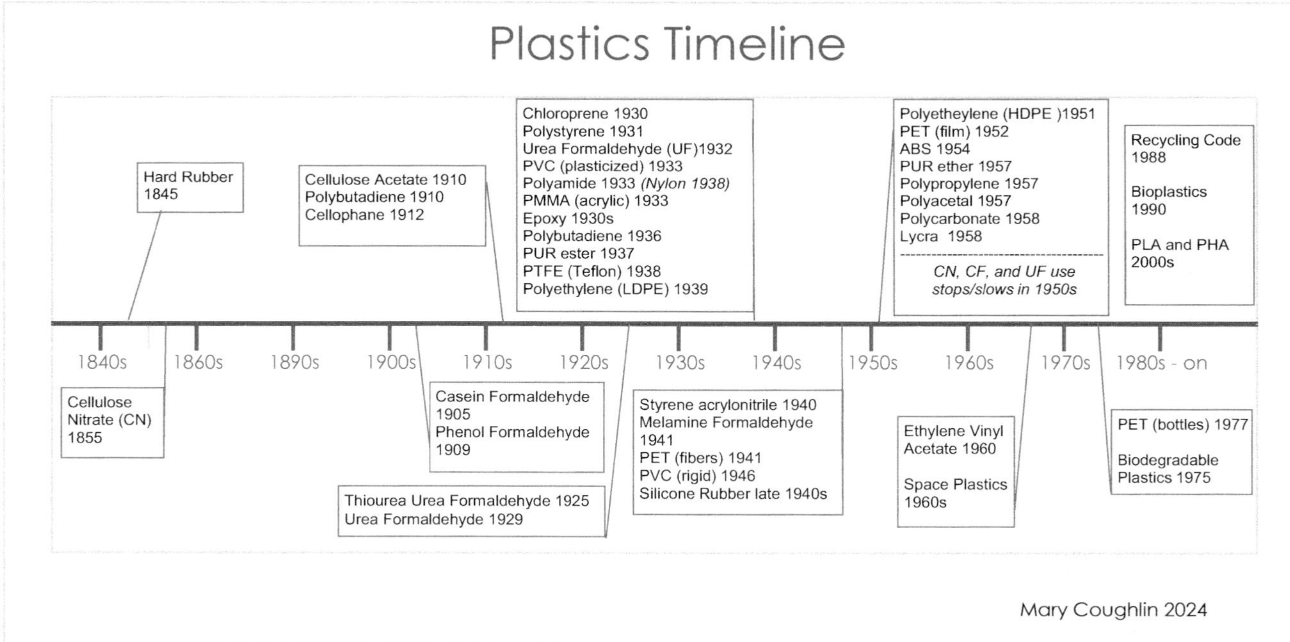

Figure 3.6 Timeline for dates when plastic was broadly manufactured, not invented. Biodegradable plastic date is for discovery (Barrett 2018). Dates may slightly differ depending on resource (MoDiP (Museum of Design in Plastics) 2007b; Plastic Identification Tool n.d.; van Oosten 2022).

with PVC (Chapter 4). So, while timelines can be very helpful, their use requires informed interpretation.

For plastic works of art, the artist, designer, or maker may have provided information about the plastic when the work was accessioned, possibly even during an interview (Sidebars 8.2 and 9.2). Information about the process, method of working, and timeframe of creation, even if not an exact date, can all help narrow down possible plastic types. Even if the artist or maker utilized commercially available plastics, they may have altered them, used them in ways unanticipated by the manufacturer, or exposed them to fabrication stresses such as solvents in paints or adhesives that could impact how that specific plastic ages which may differ from anticipated aging in an unaltered form. For plastics like epoxies that are mixed by the user, a greater variability of unknowns may exist. Even with these uncertainties, knowing as much as possible about the likely plastic or plastics that make up the final form is helpful for collection care.

Key Points for Which Plastic? Clues to Identification

The gold standard for plastic identification is scientific analysis with some instrumental techniques having the capacity to provide insight into the polymer, additives, and current condition. However, for many collections the insight gained from scientific analysis, even from relatively low-tech spot tests, is out of reach. Therefore, knowing non-analytical options to help narrow down possible plastics is useful. Knowledge gained from collection records and marks on the plastic itself, sometimes with cross-checking timelines and historical records, as well as working through identification flow charts like those found on the MoDiP website and the Plastic Identification Tool, can be invaluable for collection caretakers of all levels of experience. Because the degree of certainty of plastic identification varies based on the method or methods used, it is important to note in collection records how an identification or general classification was reached.

References

Barrett, Axel. 2018. "The History of Bioplastics." *Bioplastics News*, July 5, 2018. https://bioplasticsnews.com/2018/07/05/history-of-bioplastics/

Cucci, Costanza, Leonardo Bigazzi, and Marcello Picollo. 2013. "Fibre Optic Reflectance Spectroscopy as a Non-invasive Tool for Investigating Plastics Degradation in Contemporary Art Collections: A Methodological Study on an Expanded Polystyrene Artwork." *Journal of Cultural Heritage* 14 (4): 290–296. doi.org/10.1016/j.culher.2012.08.003

Eschner, Kat. 2018. "The Story of Brownie Wise, the Ingenious Marketer behind the Tupperware Party." *Smithsonian Magazine*, April 10, 2018. www.smithsonianmag.com/smithsonian-institution/story-brownie-wise-ingenious-marketer-behind-tupperware-party-180968658/

FilmCare.org. 2024. "Learn about Film." Rochester, NY: Image Permanence Institute. https://filmcare.org/about_film

Ishikawa, Daitaro, Yuta Hikima, and Yukihiro Ozaki. 2021. "Near-Infrared Spectroscopy and Imaging of Polymers." *Spectroscopic Techniques for Polymer Characterization: Methods, Instrumentation, Applications*: 125–164. doi.org/10.1002/9783527830312.ch5

Kelechava, Brad. 2019. "Resin Identification Codes (RICs), as Specified by ASTM D7611." American National Institute Standards, February 21, 2019. https://blog.ansi.org/2019/02/resin-identification-codes-rics-astm-d7611/

King, Rose, Josep Grau-Bové, and Katherine Curran. 2020. "Plasticiser Loss in Heritage Collections: Its Prevalence, Cause, Effect, and Methods for Analysis. *Heritage Science* 8 (123) https://doi.org/10.1186/s40494-020-00466-0

Lattuati-Derieux, Agnès, Céline Egasse, Sylvie Thao-Heu, Nathalie Balcar, Gilles Barabant, and Bertrand Lavédrine. 2013. "What Do Plastics Emit? HS-SPME-GC/MS Analyses of New Standard Plastics and Plastic Objects in Museum Collections." *Journal of Cultural Heritage* 14 (3): 238–247. https://doi.org/10.1016/j.culher.2012.06.005.

MoDiP (Museum of Design in Plastics). 2007a. "A Curator's Guide to Plastics." Arts University Bournemouth. Accessed April 19, 2024. www.modip.ac.uk/projects/curators-guide-plastics

MoDiP (Museum of Design in Plastics). 2007b. "Plastics Timeline." Arts University Bournemouth. www.modip.ac.uk/projects/curators-guide/plastics-timeline

MoDiP (Museum of Design in Plastics). 2022. "Fibre Identification." Arts University Bournemouth. www.modip.ac.uk/projects/curators-guide-synthetic-garments/curatorial-process/fibre-identification

NMAH. n.d. "Jigger." Smithsonian National Museum of American History. Accessed April 22, 2024. https://americanhistory.si.edu/collections/nmah_332580

Plastics Historical Society. n.d. "What are Plastics." Plastics Historical Society. Accessed April 19, 2024. https://plastiquarian.com/?page_id=14296

Plastic Identification Tool. n.d. "Plastics." Cultural Heritage Agency: Ministry of Education, Culture, and Science. Accessed April 19, 2024. https://plasticidentificationtool.nl/

Reffner, John A., Brooke W. Kammrath, and Samuel Kaplan. 2020. "A More Efficient Method for Synthetic Textile Fiber Analysis Using Polarized Light Microscopy." *Journal of Forensic Sciences* 65 (3): 744–750. https://doi.org/10.1111/1556-4029.14252

Richardson, Emma and Rachel Rivenc. 2012. "Hand-held FTIR." In *Preservation of Plastic Artefacts in Museum Collections*, edited by Betrand Lavédrine, Alban Fournier, and Graham Martin, 46–47. Paris: Comité des Travaux Historiques et Scientifiques.

Schilling, Michael, Rachel Rivenc, Nathalie Balcar, and Henk van Keulen. 2012. "Pyrolysis Gas Chromatography Mass Spectrometry (Py-GCMS)." In *Preservation of Plastic Artefacts in Museum Collections*, edited by Betrand Lavédrine, Alban Fournier, and Graham Martin, 61–69. Paris: Comité des Travaux Historiques et Scientifiques.

Shashoua, Yvonne. 2008. *Conservation of Plastics: Materials Science, Degradation and Preservation.* Oxford: Elsevier/Butterworth-Heinemann.

Šuštar, Vilma, Jana Kolar, Lara Lusa, Tom Learner, Michael Schilling, Rachel Rivenc, Herant Khanjian, and Dušan Koleša. 2014. "Identification of Historical Polymers Using Near-Infrared Spectroscopy." *Polymer Degradation and Stability* 107: 341–347. doi.org/10.1016/j.polymdegradstab.2013.12.035

Tupperware. 2024. "Replacement Parts." Tupperware. www.tupperware.com/pages/replacement-parts

van Oosten, Thea. 2022. *Properties of Plastics: A Guide for Conservators.* Los Angeles: Getty Conservation Institute.

van Oosten, Thea and Tom Learner. 2012. "Introduction." In *Preservation of Plastic Artefacts in Museum Collections*, edited by Betrand Lavédrine, Alban Fournier, and Graham Martin, 29–36. Paris: Comité des Travaux Historiques et Scientifiques.

Wheeler, Barbara P. and Lori J. Wilson. 2008. *Practical Forensic Microscopy: A Laboratory Manual.* West Sussex: John Wiley & Sons Ltd.

4 Problem Plastics

In 2002 R. Scott Williams of the Canadian Conservation Institute published on "Malignant Plastics" which are five plastics that can deteriorate in gravely spectacular fashion and can negatively affect nearby materials. Three of these are semi-synthetic: hard rubber (HR), cellulose nitrate (CN), and cellulose acetate (CA). The two fully synthetic plastics on this blacklist are polyvinyl chloride (PVC) and polyurethane (PUR). Other plastics not in the canon of malignant plastics can have issues as they age, sometimes the result of a bad batch or other factors, so do not get a false sense of security if a plastic is not in this grouping.

It is important to understand the terms used in the museum field to describe the visible signs of plastic degradation which are described in Table 4.1 with references made to figures throughout the book that illustrate them. Table 4.2 has the most common degradation signs for the five problem plastics that are the focus of this chapter. It is also vital to realize that deteriorating plastics can affect nearby materials such as acidic vapor corroding metal or being adsorbed by organics causing them to yellow and become brittle. The products emitted from degrading plastics can pose hazards from inhalation and contact so be sure to work in ventilated spaces while wearing nitrile gloves that are properly disposed of after use (Sidebar 2.4). Refer to Chapter 8 for considerations for recording plastic degradation in collection databases, including any insights gleaned from examination under ultraviolet (UV) light which can reveal issues that may not be apparent under normal viewing conditions (Chapter 10).

Hard Rubber: Background

HR, which is sometimes referred to as vulcanized rubber or hard vulcanized rubber, is essentially made by adding sulfur to natural rubber in quantities ranging from 5–40% in a process known as vulcanization (van Oosten 2022). The resulting thermoset plastic was usually black or brown, or occasionally red, and it was sometimes sold under the tradenames Vulcanite or Ebonite. Though discovered in 1839, its use was heaviest between 1850 and 1930 (Plastic Identification Tool n.d.). HR could be made to imitate natural materials like ebony wood, jet, and ivory, which required bleaching (Museum of Design in Plastics (MoDiP) 2007). Common uses

DOI: 10.4324/9781003489658-4

Table 4.1 Terms for Signs of Plastic Degradation

Term	*Description*	*Book Figures*
Blister	Bubbled, raised surface	8.4
Bloom	Exuded additives migrate to surface and solidify, hazy	7.4, 10.2
Break	Separation of the whole into parts; fracture or tear	2.3, 7.3, 7.4, 7.5
Brittle	Loss of strength and/or flexibility, vulnerable to breaking	2.3, 4.1, 7.2, 7.3, 7.4, 7.5, 9.1, 10.2
Channeling	Network of channel-like raised areas as a layer shrinks	–
Crack	Line or fissure that doesn't cause total breakage	10.1, 10.2
Craze	Network of cracks at surface or in body	2.3, 4.1, 7.2, 7.3, 7.5
Crumble	Friability, powdering. Starts at surface and progresses	4.8, 8.3, 9.1
Crystallization	Solidification of exuded material into 3D crystals	4.3, 5.3, 7,4, 10.2, 10.4
Delamination	Peeling away in layers	8.3
Deposits	Material on surface (e.g., dust, bloom, weeping)	2.3, 4.6, 4.7, 7.8, 10.1, 10.2, 10.3, 10.4
Discoloration	Change from original color: darkening, fading, yellowing, loss of transparency	2.3, 2.4, 2.5, 4.1, 4.2, 4.3, 4.4, 4.6, 4.7, 4.8, 5.3, 6.1, 7.2, 7.3, 7.4, 7.5, 7.8, 8.3, 9.1, 9.3, 10.3, 10.4
Flaking	Breaking off from surface in small pieces	7,4
Harden	Increased rigidity, usually from plasticizer loss	4.3, 4.7, 9.3, 10.3, 10.4
Odor	Distinctive smell; may be characteristic of plastic type	9.3
Pitting	Craters or pits in surface	–
Stain	Marks, discoloration caused by something transferred to the material rather than from the material itself	9.3
Shrinkage	Loss of volume (such as from weeping plasticizer) resulting in deformation	2.3, 7.4, 9.3, 10.1, 10.2, 10.4
Sticky/tacky	Surface stickiness caused by migration of additives	3.4, 4.2, 4.6, 5.3, 7.8, 10.3, 10.4
Warp	Concave/convex distortion from loss of volume. May look like a wave, can no longer lay flat	4.2, 4.3, 5.3, 9.3
Weeping/sweating	Additives like plasticizers migrating out, forming droplets or wet surface	4.2, 4.6, 5.2, 5.3, 7.8, 10.1, 10.3

included fountain pens, matchboxes, pipe stems, and denture palates – often tinted red with the toxic mercury containing pigment, vermillion (Neves 2023). Today, HR can still be used for musical instrument mouthpieces and bowling balls (Plastic Identification Tool n.d.). In a catalog from 1897, the HR used for "Metal Wonder Clarinets" is touted as having benefits over wood since "They are made of a special mineralized material, are as black as ebony, have a high and permanent polish, are hard, yet light and durable, are impervious to moisture, will neither swell nor warp" (Metal Wonder Clarinets 1897).

Hard Rubber: Deterioration

As HR ages it often turns brown or olive green and can dull; this is especially obvious if it was originally intended to be shiny and black, like when imitating jet or ebony. Though not considered by all to be a problem plastic, the high levels of sulfur added in manufacture can be emitted and may form sulfuric acid with moisture. The sulfur degradation products can corrode nearby metals, with silver being especially prone to tarnishing (Plastic Identification Tool n.d.).

Hard Rubber: Preventive Conservation

HR is vulnerable to photo-oxidation as well as oxidation by ambient oxygen in the surrounding environment. As found with natural rubber that was not vulcanized, storage in oxygen-free environments can be beneficial. However, this is likely not an option for most collections. The actions that can likely be taken include minimizing the duration and degree of exposure to both visible and UV light and avoiding contact with metals since copper and manganese catalyze depolymerization which means that the polymer chain breaks down as a result of this chemical interaction (Plastic Identification Tool n.d.). Additionally, sulfur from the HR itself can corrode nearby metals even if not in direct contact.

Cellulose Nitrate: Background

CN, considered to be the first commercial plastic, is a semi-synthetic plastic essentially made by soaking cellulose pulp in nitric acid and sulfuric acid (van Oosten 2022). CN is an example of a plastic whose discovery (1832) predates commercial manufacture by two decades (Plastic Identification Tool n.d.). In 1855, camphor began to be used as a stabilizer and plasticizer and wide-scale production followed over the next decades. Celluloid, patented in 1870 in the United States, grew out of a competition to replace ivory billiard balls using a mixture of ground bone, CN, and camphor (Neves 2023). Celluloid became the leading name for CN though its production is also found under other names like Parkesine, Xylonite, and nitrocellulose (MoDiP 2007). According to the Museum of Design in Plastics, popular forms included collars and cuffs; dressing table goods like combs, brushes, and mirror handles; eye glasses; jewelry; toys; and denture palates – again the desired red color sometimes coming from toxic vermillion pigment though in lower quantities than with HR since CN is easier to tint (Neves 2023). These uses generally conclude by WWII. (MoDiP 2007). CN was also made into fibers to imitate silk and was marketed as Chardonnet silk and later as Fibersilk (Conservation and Art Materials Encyclopedia Online (CAMEO) 2022).

At first CN was used to imitate naturally derived luxury goods like tortoiseshell and ivory, with some imitation CN ivory sold under names like French Ivory and Ivoride. But just because CN was made to imitate a product like ivory, it did not mean that they were intended it to be precious. Even at this early stage of plastics history, some CN ivory products were intended as expendable novelty ephemera. CN's versatility meant that its production expanded beyond imitating nature to become sheets and three-dimensional forms in a kaleidoscope of colors. Since CN can also be transparent, it was used as coatings, lacquers, sheets, and as film for photographs and motion pictures. Its transparency and other working properties made CN popular for cartoon cel animation (cel shortened from the tradename Celluloid).

It is important to remember that CN was also used as an explosive and in this form is typically called nitrocellulose or gun cotton. The fact that CN could be used as an explosive means that it is flammable and this property led to its decline in use by 1950 (Plastic Identification Tool n.d.). However, most of the CN flammability concerns for collection care are with film formats because of its formulation and bulk storage. Even at the time of use, combustion was a concern with countless cinema and film studio fires in the early 20th century leading to the development of CA Safety Film as a replacement in 1910 (Plastic Identification Tool n.d.).

Though production of CN dwindled by 1950, old stock, particularly for cel animation, continued to be used. Today CN is still used for ping pong balls and lacquers (van Oosten 2022). A CN lacquer that may be familiar in collection care is Agateen, which is used as a coating to slow silver tarnishing.

As a reminder, Nitrate Test papers and the diphenylamine spot test are relatively easy ways to determine if a material is made of CN (Chapter 3).

Cellulose Nitrate: Deterioration

"The stability of cellulose nitrate is strongly influenced by the amount of nitrogen present in it: the greater the nitrogen content, the more unstable the product" (Williams 1994). Not surprisingly, explosives are the CN products with the most nitrogen, next are films, followed by some lacquers and then three-dimensional forms, making the latter more stable though they can still pose issues as they age.

Acids added during production also affect CN degradation. If sulfuric acid is not fully rinsed away, it can catalyze deterioration. Additionally, lignin in cellulose has the potential to be an additional acidic source. Humidity also plays an important role in aging because CN is vulnerable to hydrolysis, the process in which water breaks chemical bonds, which causes CN to emit reactive volatile nitrogen oxides which can be converted to nitric acid by reaction with moisture at levels at or above about 50% Relative Humidity (RH) (van Oosten 2005). Emitted nitric acid not only degrades CN, it can also affect other materials. Those in charge of monitoring storage and exhibition environments, know that in many places for much of the year, maintaining a stable RH below 50% is difficult, often requiring specialized equipment that uses a lot of energy and may not always be feasible or appropriate. That said, because of CN's vulnerability, it is important to monitor RH and try to mitigate it (Chapter 5).

Depending on the final CN form, additives such as plasticizers and colorants may have been added. And, as with all plastics, these additives can affect how CN ages. Camphor was a common plasticizer for CN through the 1930s but is not well bonded and at room temperature sublimates, meaning that it goes from a solid to a vapor. As camphor vaporizes, small holes form that allow atmospheric water to penetrate and form nitric acid below the surface which accelerates degradation within the CN object. In the late 1930s, other plasticizers such as phthalates and triphenyl phosphate replace camphor. But regardless of the type, plasticizers play a protective stabilizing role for CN. When they weep out of the polymer, internal crizzling forms and chemical degradation accelerates within (Shashoua 2008). Eventually, complete breakage can occur and the plastic falls apart. Because CN degradation can start below the surface, objects may look intact but actually be further along their degradation pathway. CN products intended to mimic ivory often used zinc oxide as a pigment and filler. The zinc can neutralize nitric acid to more stable zinc nitrates, thereby slowing degradation (Derrick, Stulik, and Ordonez 1993). Like sunscreen, zinc oxide also likely shields these CN objects from some light-induced degradation (Fenn 1993). However, over time zinc oxide can transform to acidic zinc oxalate and if noted during scientific analysis is a degradation marker (Quye et al 2011).

Telltale signs of CN degradation are cube-like fissures, yellowing, crazing, warping, weeping, cracking, and sometimes the formation of whitish crystals from additives like plasticizers that migrated to the surface. Emitted acidic vapor can have an acrid smell, corrode nearby metals, and be adsorbed by organics, like paper, which can cause them to yellow and become brittle. Acidic off-gassing accelerates degradation of the CN itself meaning that the process is autocatalytic or self-perpetuating so it is best to not let acidic vapor build-up. Thus, ventilation is a good default. For thick objects, like the one in Figure 4.1, it takes more time for the nitrogen oxide gases to diffuse out. Since these reactive gasses are trapped for longer periods of time, more acid can form and fuel an internal autocatalytic cycle of degradation (Williams 1994). Methods for determining if CN is off-gassing acids are covered in Chapter 7.

Cellulose Nitrate: Preventive Conservation

To counter the degradation mechanism hydrolysis, CN collections are ideally stored at 20–30% RH ±5% with as minimal fluctuations as possible (Plastic Identification Tool n.d.). Because CN degradation is autocatalytic, CN should not be kept in a sealed environment that traps acidic vapor, such as in a polyethylene (PE) bag. Instead, ventilate. However, the threat of acidic off-gassing impacting nearby materials is a concern and metals can corrode and organics like paper can

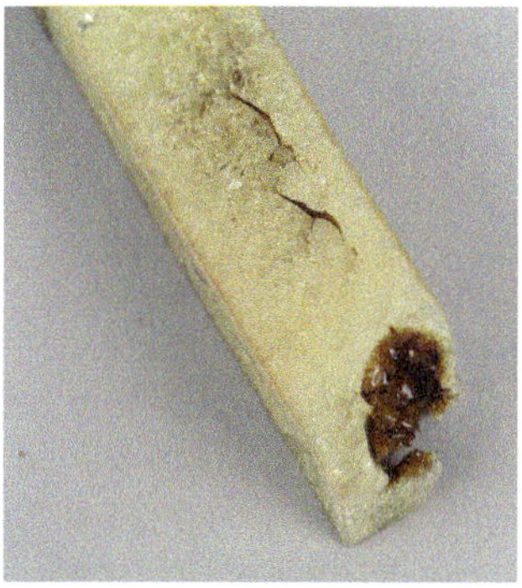

Figure 4.1 Discoloration, crazing, and embrittlement in thick CN sample from 1942. NMAH 1983.0538.87. Photographed by author, Division of Medicine and Science, National Museum of American History, Smithsonian Institution.

adsorb the vapor causing them to become acidic. Some of the concern about cross-contamination can be lessened by monitoring how far the acidic vapor is spreading and arranging storage accordingly (Chapter 7).

For CN objects that are actively off-gassing, there may be a role for adsorbents, like activated charcoal/carbon, to play in adsorbing acidic vapor to slow autocatalytic degradation and museums have utilized this approach since the 1990s (Shashoua 2014). However, research on CA has shown that adsorbents, like activated charcoal/carbon and zeolites, may be too aggressive for plastics and can actively pull out plasticizers and fire retardants so adsorbents cannot be used in every situation but may have a role if plasticizers are already lost but the plastic is still off-gassing acids (Shashoua 2013). Buffered archival papers and corrugated boards, like blueboard, have been shown to be effective in countering acidic off-gassing and have the added benefit of already being familiar products in collection storage (Shashoua 2014). For those unfamiliar with blueboard (also called Blue board, among other names), it is an acid and lignin-free corrugated archival board that is usually buffered with 3% calcium carbonate to pH of 7.5–9.5 (CAMEO 2022). It is usually light blue, hence the name, but archival corrugated boards with these properties can also be gray, tan, or white.

Storing at cooler temperatures slows down chemical reactions. For CN film, low-temperature storage is important in order to keep temperatures below the autoignition point, which for new CN is 150°C (302°F) but drops to 50°C (122°F) in severely degraded CN film (Williams 1994). Heavily deteriorated CN film can spontaneously combust if stored at temperatures slightly above 38°C (100°F) for a prolonged period (Kodak 2024). For museum collections, the concerns of CN fire are for film storage rather than three-dimensional objects because film is more nitrated and may be stored packed together in great quantities with inadequate ventilation allowing acidic vapor to build-up. So, if ignition temperatures are reached there is a lot of fuel for the fire. To counter this risk as well as to slow chemical degradation, CN film is best stored frozen (O°C or 32°F), especially if actively degrading, or at cold temperatures (4°C or 40°F) with both temperature ranges assuming RH of 30–50% (Image Permanence Institute (IPI) 2024). Because hydrolysis plays a role in degradation, storing CN film below 40% RH curbs its impact but the film will become more brittle which introduces risks when handling (Kodak 2024). The National Film and Sound Archive of Australia notes that a reduction in temperature by 6°C results in about 50% less nitrogen dioxide being emitted from CN film. For low-temperature storage of any range, it is vital to have a back-up generator in case of power loss so that temperatures do not rise dangerously high. If lowering the temperature in storage is not possible for CN film, keep temperatures below 21°C (70°F), store CN film in smaller groupings in proper containers, and ventilate (Kodak 2024). Because of CN film fire risks, rather than retaining film, duplication and disposal can be prudent collection care measures.

Even if you do not have CN film collections, that does not mean you can ignore temperature for CN objects. The glass transition temperature (Tg), or the temperature at which the plastic shifts from hard and rigid to flexible and soft, for CN is 53–66°C (about 127–150°F) (Plastic Identification Tool n.d.). Storage in warm climates during summer in spaces that lack climate control and get hot, like attics, could reach these softening temperatures and affect this thermoplastic material. So, regardless of the form, storing CN in as cold temperature as possible, while maintaining an RH below 50% but not lower than 20%, is advisable.

Cellulose Acetate: Background

CA is another example of a plastic whose discovery predates widespread use, in this case by about 50 years. CA was discovered in 1865 and altered to create artificial silk called viscose in 1892, which in 1904 became known as rayon (MoDiP 2007). It was not until 1904 that CA objects begin to be made and broad commercial production starts in 1910, which is why that is the date on the timeline in Figure 3.6 (MoDiP Timeline 2007; Plastic Identification Tool n.d.). Just like its predecessor, CN, CA could be transparent as well as made into a variety of translucent and opaque colors and take many forms including as film, coatings, sheets, fibers, and three-dimensional objects. CA marketing often emphasized its versatility and that it did not have the same fire risk as CN, which was especially important for CA Safety Film which was developed to counter CN film fire concerns.

Use of CA grew in the late 1920s and ebbed by the late 1960s (MoDiP 2007). During that period besides for film, CA was also used for acetate rayon fabrics for clothing and lampshades and for rigid objects like eyeglass frames, typewriter keys, cigarette cases, brush handles, and for toys, including Lego bricks between 1949–1963 (Plastic Identification Tool n.d.). It was manufactured under tradenames like Celanese, Estron, Tenite, Lumarith, and Safety Film.

Today, interest in CA is renewed because of the appeal of it breaking down to products that are advertised as less harmful to the environment (see CA Deterioration section below). Eyeglass manufacturers in particular are re-embracing CA, just as they did in the mid-20th century (Figure 7.4). A quick internet search will show CA eyeglass advertisements emphasizing that CA is a plant-based plastic from renewable resources, is hypoallergenic, lightweight, strong, durable, and available in an endless variety of colors and designs. It is fascinating to compare the similarities between contemporary CA ads and those from the early decades of its use, though today emphasis is on assuaging environmental rather than flammability concerns.

Cellulose Acetate: Deterioration

CA is similar to CN since it's base is also cellulose pulp and sulfuric acid but instead of nitric acid, CA is made with acetic acid (vinegar). And as with CN, if the sulfuric acid remains after processing, it accelerates CA deterioration. When CA is far along on its degradation pathway, to the point that the polymer itself is degrading or depolymerizing, CA will break down into these components which is why CA is touted as a more sustainable plastic. However, this assertion disregards additives in the manufacturing process and these can be much more environmentally harmful than cellulose and vinegar (Sidebar 2.4).

A familiar indicator of CA degradation is a vinegar smell as it forms acetic acid at RH levels at or above about 50% (van Oosten 2005). Another common degradation sign is loss of plasticizers resulting in weeping, tacky surfaces as the plasticizers migrate out and may form acidic compounds (Figure 4.2). The loss of plasticizers can cause CA to distort and warp as it loses mass (Figure 4.2) and can lead to cracks, breakage, and rigidity as CA becomes less and less plastic (Figure 4.3). Sometimes the additives that exude from CA will form white crystals on the surface, which was the case within the curled area in Figure 4.3. If the acidic vapor or acidic weeping interacts with other materials, such as metals, it can cause their deterioration as well (Figure 2.3).

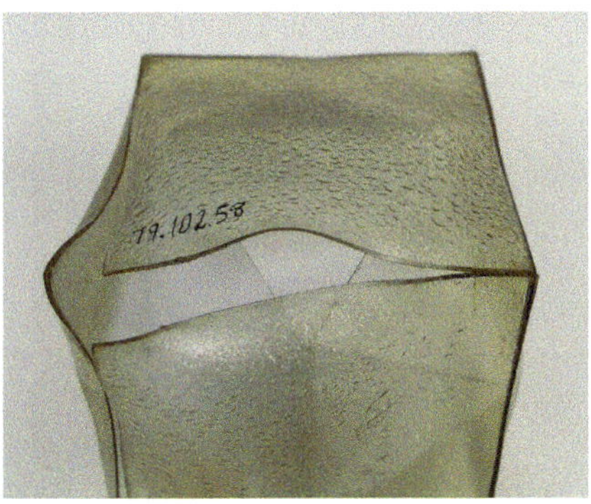

Figure 4.2 Circa 1930s geometric model made of a CN internal form and CA external form. CA is yellowed, weeping, and warped from plasticizer migration. Once in the NMAH mathematics collections (1979.0102.058). Photographed by author, Division of Medicine and Science, National Museum of American History, Smithsonian Institution.

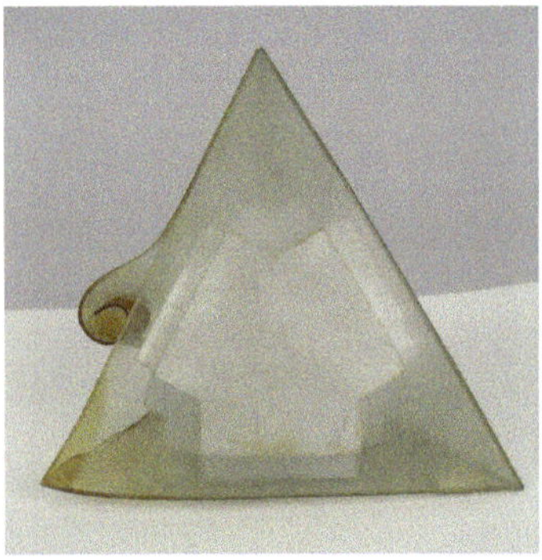

Figure 4.3 Circa 1930s geometric model made of a CN internal form and CA external form. CA is yellowed and warped; curled section is rigid from plasticizer loss, some of which crystallized. Once in the NMAH mathematics collections (1979.0102.312). Photographed by author, Division of Medicine and Science, National Museum of American History, Smithsonian Institution.

Cellulose Acetate: Preventive Conservation

Like CN, CA deterioration is driven by hydrolysis so it is vital to monitor and mitigate RH, keeping it below 50%. Aiming for 20–30% RH ±5% with as minimal fluctuations as possible is a good target range (Plastic Identification Tool n.d.). Hydrolysis will drive the release of acetic acid which will further accelerate degradation in an autocatalytic cycle so, as with CN, default to ventilation though there may be times when the use of scavengers is beneficial (Townsend et al 2023). Though this comes with the warning that adsorbents like activated carbon/charcoal may actively pull out plasticizers and fire retardants (Shashoua 2014). Chapter 7 outlines some options for monitoring for acidic off-gassing with A-D Strips being specifically made to monitor CA film. Buffered archival papers and corrugated boards, like blueboard, are again a good choice since they counter-released acids without promoting plasticizer migration.

The degradation of CA will be slowed by storing in as low temperatures as possible but RH cannot be ignored. The Image Permanence Institute (IPI) notes that "…acetate film will begin to seriously degrade in about 50 years at room temperature and moderate RH. Periods of higher temperature and dampness will accelerate the process, and cold and dry periods will slow it down" (2022). CA film should be kept in frozen storage (O°C or 32°F) if actively deteriorating, or kept in cold temperatures (4°C or 40°F) if currently in stable condition. Both frozen and cold storage assumes an RH in the 30–50% range (IPI 2024).

Cellulose Nitrate and Cellulose Acetate

As outlined above, CN and CA have many similarities since they are both chemically cellulose esters, however, this does not mean they are compatible together.

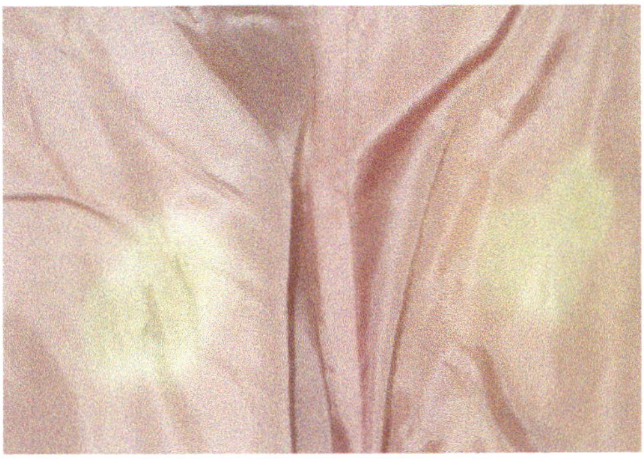

Figure 4.4 Fume fading of acetate fabric caused by contact with CN buttons.

Given their overlap in time periods and the myriad forms they can take, it is not unheard of that these two types of plastics are found in a single object and their degradation products can negatively affect each other. In fact, the objects in Figures 4.2 and 4.3 have CA exterior forms encasing CN interior forms, which did not do each other any favors. An example of fume fading on acetate fabric caused by contact with CN buttons is seen in Figure 4.4 and is the reaction that the protective measure in Sidebar 4.1 is trying to avoid! In the past, acetate fabric has been used as a test fabric to indicate degrading plastics whose off-gassing can cause the CA fabric to change color with red, green, and purple typically being the most sensitive colors (Fenn 2005).

If you are unsure if you have CN or CA but suspect one or both of them are degrading, there are test strips discussed in Chapter 7 that can aid in monitoring and, in some cases, can be used to help distinguish between the two plastics (Figures 7.4 and 7.5). If you are still uncertain if something is CN or CA, thankfully, the environmental recommendations and material choices for storage and display are the same.

Sidebar 4.1 CN and CA Buttons

Christopher Mazza

At The Costume Institute, Fourier transform infrared (FTIR) analysis has identified the wide use of the malignant plastics CN and CA, often in combination, for the manufacture of buttons from the 1930s to the early 1950s. As costume is often a composite object, these early plastics have the potential to damage associated textile materials. For a 1942 Vera Maxwell cotton twill uniform (C.I.47.60.1), the deteriorating buttons (a CA substrate with a CN coating) developed multiple internal cracks and the cotton twill in direct contact with the buttons fully degraded while the cotton twill in the immediate vicinity underwent significant bleaching. To prevent this type of damage, it is the protocol at The Costume Institute to remove (and retain separately) buttons verified as containing either one of these malignant plastics. A more general procedure, if a button's material has the potential to impact the condition of the associated textile, is to minimize contact by inserting an archival barrier, such as silicone release paper, as seen in Figure 4.5.

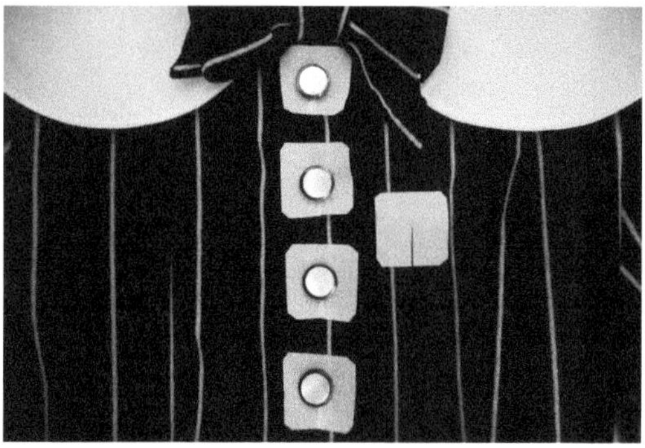

Figure 4.5 Mockup of buttons with silicone release paper barrier. Credit: Christopher Mazza.

Polyvinyl Chloride: Background

Polyvinyl chloride (PVC), which is also written as poly (vinyl chloride), has a winding path from discovery in Germany in 1872 to manufacturing in the 1930s that includes overlapping experimentations in different countries with references citing slightly differing dates for discovery, development, and manufacture. From the 1930s–1960s, PVC was made by reacting chlorine and acetylene. Beginning in the 1960s and continuing today, PVC is manufactured using ethylene from fossil fuels and chlorine to produce vinyl chloride monomers that are then reacted to form the synthetic polymer, PVC.

PVC was technically a late-19th-century discovery but its production was difficult which slowed its transition to a marketable material. Experimentation and increased efforts in production occurred during WWI but PVC remained problematic to manufacture until the 1930s when new plasticizers began to be produced. The 1940s saw increased use of PVC, helped by wartime needs such as using PVC as a rubber substitute. When PVC first went to market, consumers complained of PVC smelling, sweating, and becoming brittle (which are similar to issues with degrading PVC). In 1944, the US Society of the Plastics Industries decided to educate the public about how to care for and appreciate plastics using informative labels (Mulder and Knot 2001). In the early 1950s, PVC production grew by about 15% a year as PVC products and manufacturing became global (Mulder and Knot 2001). Despite an initially rocky acceptance by consumers, during the 1950s PVC production steadily grew and it remains one of the most produced plastics today.

It is important to note that PVC has two primary forms: plasticized and unplasticized. As should now be obvious, plasticized PVC has added plasticizers to impart flexibility with anywhere from 15–50% of the total weight as plasticizers (Shashoua 2008). As outlined in the deterioration section below, it is the plasticizers in PVC that tend to cause the most obvious signs of degradation. Plasticized PVC is used to make shower curtains, toys like Barbie dolls (including sometimes dolls' hair), inflatables like balls and floatation toys, transparent film, insulation for electrical wires, and in medical applications for blood bags and flexible tubing. Plasticized PVC can also be used for coatings including on textiles to imitate leather.

Unplasticized PVC production did not begin until 1946 with toys being among the first products (Plastic Identification Tool n.d.). Because of its rigidity, it is well-suited for applications like pipes and window frames. In the late 1940s/early 1950s, PVC was made into a copolymer with polyvinyl acetate (PVAc) to make vinyl records (Plastic Identification Tool n.d.; van Oosten 2022).

Polyvinyl Chloride: Deterioration

One of the factors that influences the behavior of PVC as a final product, as well as how it ages and the problems it causes in collections, is the addition of plasticizers. Plasticizers are weakly bonded to the PVC polymer and can migrate out leaving the surface sticky, greasy, and discolored, sometimes forming a whitish bloom. The tacky surfaces are prone to trapping dust and other pollutants. Remember that other additives can migrate out as well so you cannot assume that every weeping material or surface bloom is plasticizer but regardless of the source, these are signs of deterioration.

The main degradation mechanism for PVC is called dehydrochlorination. During this process, hydrogen chloride is released and can form hydrochloric acid with atmospheric water thus setting up an autocatalytic reaction of self-perpetuating degradation (Shashoua 2008). Over time the polymer becomes less amorphous and this increased crystallinity pushes more plasticizer out (Ledoux et al 2023). As plasticizers are lost, the plastic loses mass, can start to distort, becomes less flexible and, ultimately, rigid and prone to cracking. Environmental stressors like high temperature and UV light can promote dehydrochlorination thus accelerating degradation (Williams 2019). As PVC deteriorates it absorbs light in the violet, blue, and green spectrum which explains some discoloration from yellow through to orange and red and finally brown and black as degradation progresses (Shashoua 2008).

Though a myriad of plasticizers has been available to manufactures, phthalates have been the most frequently used with di-(2-ethylhexyl) phthalate (DEHP) the most commonly used plasticizer in PVC since the mid-20th century. DEHP itself introduces degradation issues for collections because DEHP molecules can form phthalic acid as they degrade. This change is not often observable until it is very degraded, as evidenced by a strong odor and discoloration (Shashoua 2008). In the late 1960s, concerns of leaching plasticizers on human health were raised but it was not until the 21st century that DEHP and bisphenol A (BPA) (an additive sometimes found in PVC and other plastics) began to be banned from certain products in some countries (Halden 2010). Because of these potential health risks, take protective action when handling plastics (Sidebar 2.4). If gloves become wet from handling weeping plastic, immediately remove them, wash hands, and replace with a new pair of nitrile gloves (Ledoux 2023). Remember because of health and environmental concerns, these used gloves are now contaminated and should be considered hazardous waste.

Since most of the issues with PVC are a result of it being plasticized, unplasticized PVC is not considered a problem plastic (Plastic Identification Tool n.d.). However, both plasticized and unplasticized PVC are prone to yellowing and darkening over time which is often the visible result of dehydrochlorination and the formation of acids poses risks to nearby materials (King, Grau-Bové, and Curran 2020). Stabilizers added during manufacture slow this process but they become used up over time. The inclusion of stabilizers means that manufacturers know that PVC essentially has lifespans of use and rarely does industry expect this use to exceed 20 years (Shashoua 2008). In collection surveys, most plasticized PVC starts to visibly degrade after 5–35 years with obvious variables coming from different formulations, stresses from use and environmental conditions, and the length of time from manufacture to acquisition (Shashoua 2008). What the finite lifespan of PVC, and other problem plastics, mean for collection care is explored in Chapter 9.

Polyvinyl Chloride: Preventive Conservation

As discussed in Chapter 3, scientific analysis with FTIR or Raman spectroscopy could provide information about the PVC polymer and additives that would give more insight into current condition. If the need or capacity is to only determine if an object is PVC, both plasticized and unplasticized PVC can be identified with the Beilstein Test or through the detection of chlorine with X-ray fluorescence (XRF) (Chapter 3). Degrading plasticized PVC seems to cause a reaction with Ozone Test sticks which may be helpful monitors (Chapter 7).

Though both plasticized and unplasticized PVC are made from the same polymer, it is the difference in additives, namely plasticizers, that dictate different preventive conservation recommendations. Issues for plasticized PVC stem from loss of plasticizers so care should be taken to avoid leaching out those additives. It can be surprising to think of some plastics as adsorbent but PE, which is ubiquitous in collection storage as bags and sheeting, can not only adsorb deterioration products, as discussed in the storage of deteriorating CN in Chapter 3, but can also actively pull out plasticizers. Scavengers like zeolites can also accelerate plasticizer loss so neither should be used with plasticized PVC collections (Shashoua 2003). One research team determined in accelerated aging tests that when polyethylene terephthalate (PET or PETE), which is a form of polyester used to make products familiar in collection care under the names Mylar or Melinex, was in contact with PVC, the PVC discolored at contact points. Though PET did not cause plasticizer migration from PVC it seemed to contribute to the degradation of the plasticizer itself as noted by the formation of phthalic acid (Royaux et al 2020). That said, other researchers found wrapping PVC in PET to have benefits, particularly in cool storage (Sidebar 5.1).

Since a risk for PVC is leaching out plasticizers, studies have shown that there is an advantage to storing plasticized PVC in non-adsorbent containers, such as glass, because an equilibrium between the PVC and the environment can be reached slowing the migration of plasticizers (Sidebar 4.2) (Shashoua 2003). It also prevents dust from sticking to tacky surfaces and protects other materials from off-gassing acids released as PVC degrades (Ledoux et al 2023). If PVC is stored in non-adsorbent containers, monitoring with Ozone Test sticks and/or A-D Strips, as discussed in Chapter 7, may be insightful for understanding current condition.

Storing in low temperatures, as discussed in Chapter 5, will slow chemical reactions and plasticizer migration. However, it can come with some risks namely increased embrittlement that can result in physical damage, especially in composite objects with materials that may contract at different rates upon cooling and expand differently when brought back to room temperature. So, it is important to acclimate in and out of low temperatures as well as taking measure to avoid condensation. One study of PVC dolls identified whitish blooms as a lubricant used in the manufacturing process that worked its way out when stored at lower temperatures. Despite the lubricant bloom, collection stewards decided to keep the PVC in low temperature storage in order to retard plasticizer migration since its loss would be fundamentally destructive (Lauridsen et al 2017). Conservators at Harvard Art Museums found that PVC sealed in Mylar/Melinex in cool storage actually helped migrated plasticizers work back in (Sidebar 5.1).

Since unplasticized PVC is not vulnerable to damage from plasticizer loss, the use of adsorbents like zeolites can actually be beneficial to counter acidic off-gassing (Plastic Identification Tool n.d.). If acidic vapor is suspected for PVC, refer to Chapter 7 for monitoring options.

Because light is a catalyst for some of the degradation pathways for PVC, it is best to keep all types of PVC at low visible light levels and avoid UV light exposure as much as possible.

Sidebar 4.2 Comparison of Storage Impact for Plasticized PVC Watches

Swatch watches known for their clear or brightly designed plastic watch faces and bands were an international fad of the 1980s. Swatch's Jelly Fish watch, introduced in 1983, was the first transparent watch and had a clear case showing the internal watch mechanisms with a translucent band (Farago 2019). It remained a popular model for the company and is important in design history with examples in the collections of the Museum of Modern Art (MoMA), Cooper Hewitt Smithsonian Design Museum, and the National Museum of American History (NMAH), among other museums.

The NMAH watch entered the collection in 1996, as indicated by its accession number, and by the time of an inspection in 2005, roughly 20 years after the likely date of manufacture, the once clear rigid mount for the watch face yellowed as had the watchband. The band was also tacky, likely caused by the migration of plasticizers but remained flexible (Figure 4.6). Another conservator working in the NMAH Objects Conservation Lab at this time had a similar watch from her 16th birthday in January 1987. Its watch face mount and band also yellowed but the band was rigid and locked into a curved shape (Figure 4.7). Why did these two similar watches both yellow but differ in flexibility?

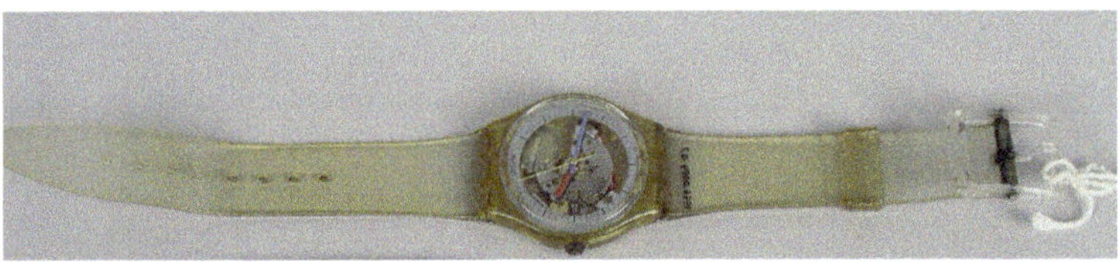

Figure 4.6 Swatch watch with tacky, yellowed band and watch face mount. Stored flat in open storage. NMAH 1996.0089.07. Photographed by author, Division of Work and Industry, National Museum of American History, Smithsonian Institution.

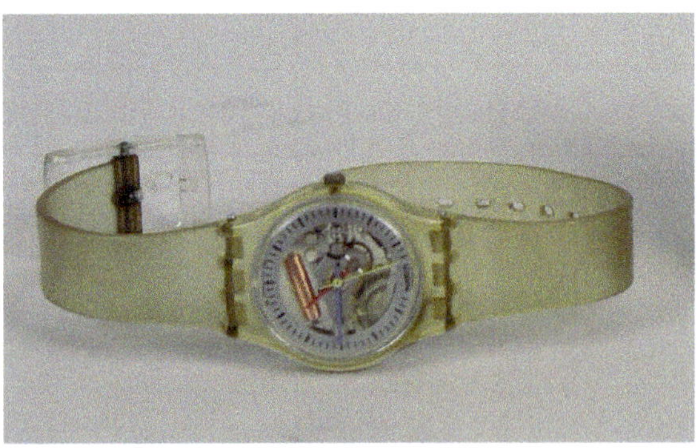

Figure 4.7 Swatch watch with yellowed band and watch face mount. Stored curved in sealed rigid plastic container.

To answer that question, the first step was to identify the plastic used on each watchband. Utilizing the Beilstein Test, as discussed in Chapter 3, a vivid green flame formed for each band, indicating chlorine. The bands were also analyzed with XRF which also detected chlorine. The testing confirmation of chlorine combined with the date of manufacture, mass production status, and the need to be flexible all indicate that both bands are PVC. The MoDiP timeline also marks the introduction of Swatch watches in 1983 with PVC bands (2007).

Discoloration, in this case yellowing, is a common sign of PVC degradation as is plasticizer loss. The main difference to the aging of the bands had to do with storage. It is unknown how the NMAH watch was stored between manufacture and accessioning into the museum but from 1996–2005 it was kept flat in a paper tray on open shelves. This open storage for about a decade likely increased the rate of plasticizer loss since it could never reach an equilibrium with its surroundings. One study found that PVC stored in ventilated storage in standard museum conditions lost more than three times the amount of plasticizer compared to PVC stored sealed at low temperatures (Ledoux et al 2023). Additionally, placement on open shelving allowed for dust to deposit and stick to the tacky surface. The birthday watch was stored mounted as if wrapped around a wrist in an aftermarket clear case made of an unknown rigid plastic. The MoDIP timeline indicates that original Swatch cases are acrylonitrile butadiene styrene (ABS) and are flat (2007). Despite being in an enclosed environment, this gifted watch's band was rigid, likely from plasticizer loss, and could not lay flat. It is likely that the rigid plastic case leached out plasticizers from the band.

An effort was made to try and compare the degree of degradation of the bands. FTIR analysis of surface samples from each band indicated the presence of phthalate plasticizers, confirming the migration of plasticizers out of the polymer. Both watchbands had acidic surfaces of pH 5, likely from the formation of acidic degradation products. Together, these analyses confirmed that each watch was far along on its degradation pathway.

Next the impact to changes in storage were studied. First, each watch was weighed then each watchband wiped clean. After surface cleaning, each weighed 0.01 gram less showing that material was removed. The NMAH watch was then placed in a sealed glass container for five months. After that time, it was still flexible but stiffer than it used to be. It was slightly tacky but nowhere near the extent of tackiness it had before cleaning and storing in the covered glass container, which may show a slowing of plasticizer migration. Of note is that the ends of the watch which had previously been curled up from being kept in a storage tray that was too small now relaxed and laid flat. The gifted watch was kept in open storage for the five-month test period and after that time, the watch was much more flexible, slightly tacky, and could now be made to lay flat though would not stay flat on its own.

Based on this small experiment, storing PVC in a non-adsorbent closed container does seem to slow some of the plasticizer weeping, as indicated by the NMAH watch. It is likely that the birthday watch became more flexible in open storage because of increased migration of the remaining plasticizer but keeping the watch in open storage would likely lead to complete rigidity and eventual failure of the PVC band. Therefore, it was recommended that both watches be kept in sealed glass containers.

Polyurethane: Background

Polyurethane (PU or PUR) is a truly dynamic plastic that can take many forms. On the Plastic Identification Tool website, PUR is listed under five different categories (Plastic Identification Tool n.d.). PUR was discovered in 1937 in Germany but WWII delayed broader manufacturing. However, wartime shortages of rubber meant that PUR in both its rubbery form as well as for coatings for airplane finishes were manufactured. After WWII, an increase in variety of forms enter the marketplace over time:

- Foam (ester): 1937
- Coatings: 1940s
- Adhesives: 1950s
- Synthetic leather: 1953
- Foam (ether): 1957
- Spray foam: 1957 (ester) 1960 (ether)
- Fibers (Spandex, Lycra, Elastane): 1958
- Imitation wood: 1970
- Artificial heart: 1980
- Memory foam mattress: 1991
- Expanded PUR foam used in sneakers: circa 2010

Though all PUR has the same polymer structure, changes to its formulations mean that PUR is an elastomer that can behave like a thermoplastic or a thermoset. Thermoplastic PUR forms are used for coatings, adhesives, sealants, and fibers. Thermoset PUR is primarily used as either flexible or rigid foams. Flexible foams are made with a blowing agent such as carbon dioxide; the more blowing agent used, the softer the foam since the bubbles create a more open cell structure making them ideal for cushions and mattresses. Rigid foams are made by trapping gas in the foam cells. This less open matrix makes these foams more rigid and contributes to their sound and heat-insulating properties. The vast majority of PUR production today is for foams. PUR foams can be found in collections as accessioned objects and artworks or as storage, transport, or display supports, which is ill-advised for long-term use (Chapter 6).

Soft PUR foams can be further categorized as either poly*ester* polyurethane (PUR ester) or poly*ether* polyurethane (PUR ether). The PUR ester foams have slightly smaller cells and are often dark in color. They are used less often today but may still be used for sound insulation. PUR ether foams are now more common and available in a range of colors.

Polyurethane: Deterioration

While not always possible to distinguish between polyester and polyether PUR foams without scientific analysis (Chapter 3), they have different vulnerabilities and degradation pathways that are important to understand.

The ester form is most vulnerable to hydrolysis, degradation catalyzed by high RH. Acidic degradation products can be produced which promote further degradation. Once degradation starts the process is autocatalytic and quick. Or in other words, it can seem like the foam is fine then all of a sudden appears discolored, tacky, loses its ability to bounce back to shape, and crumbles.

Polyurethane ether foam degrades primarily by photo-oxidation which is driven by exposure to oxygen and light. It typically manifests as a yellowed or discolored surface that crumbles and stops bouncing back if force is applied and then removed, as seen in Figure 4.8.

The large surface area of both types of foams make them vulnerable since they have so many exposed sights for degradation reactions to begin. This means that deterioration may be most obvious at the surface but can be occurring in the interior as well. If air was blown through to make the foam, it creates a built-in opportunity

(a) (b)

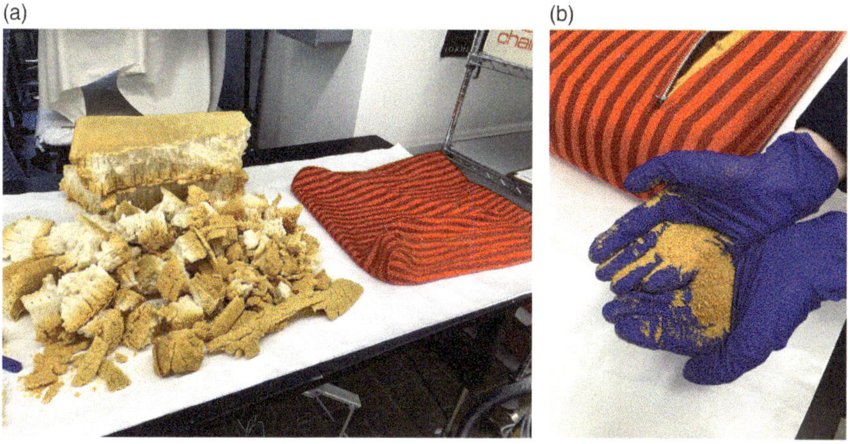

Figure 4.8 Polyurethane foam upholstery replacement during treatment of an Alexander Girard chair for Braniff Airways, treated by Kress Fellow Geena Roth. Note darker discolored surfaces in (a) that are crumbling into powder (b). Credit: Jessica Walthew for Cooper Hewitt Smithsonian Design Museum.

for oxidation reactions (Shashoua 2008). Coated or painted foams may age better since they have a barrier between the environment and the foam itself but even when not directly exposed, PUR foam will still degrade, as can occur in the thin PUR linings of baseball hats and costumes. The assumed "useful lifetime" for non-coated foam is about 20–25 years but issues like yellowing and crumbling can start sooner (van Oosten 2011).

Additives like mold release agents and surfactants used during manufacture can cause the release of volatile organic compounds (VOCs) from PUR foam (Hornyák-Mester et al 2023). The effect that these VOCs have on the indoor environment both in terms of human health and potential negative interaction with other materials is something to be aware of with PUR foams.

Non-foam PUR, like coatings and fibers, tend to age better because they do not have the high number of reactive sites that foams do. That said, coatings and sealants will fail over time and once stretchy PUR fibers will eventually stop bouncing back into shape.

Polyurethane: Preventive Conservation

Since polyester and polyether PUR foams have different environmental sensitivities, ideally, they should be identified to create targeted storage and exhibition environments. There are various scientific instrumental techniques discussed in Chapter 3 that are available to distinguish between the forms, namely FTIR spectroscopy, Py-GC/MS, and HS-SPME-GC/MS. HS-SPME-GC/MS has the advantage of being completely non-invasive since it is testing the volatiles around the foam rather than the foam itself. Researchers using this method are able to distinguish between ether and ester PUR "at any point of their degradations" (Lattuati-Derieux, Thao-Heu, and Lavédrine 2011).

If analytical instrumentation is not available, another option that is more in line with spot test approaches exists but requires access to a fume hood, proper personal protective equipment, chemicals, and the ability to take a 1 mm^2 sample of the foam which is placed in a 10% potassium hydroxide (KOH) solution. KOH is a strong base which makes it caustic so safety precautions need to be taken. If the foam sample dissolves, it is a PUR ester. If it does not, it is the ether form (van Oosten 2011).

Since PUR ester foams are vulnerable to hydrolysis, RH should be kept between 10–30% (Plastic Identification Tool n.d.). Lowering the temperature will slow down hydrolysis as well but since temperature and RH are inverse to each other, make sure that lowering the temperature does not raise the RH.

PUR ether foams degrade by photo-oxidation so storing in an oxygen-free (anoxic) environment would be ideal though not usually feasible. If an anoxic environment is not possible, keep at low light levels and at lower temperatures to slow the rate of degradation. However, since past exposure may have catalyzed degradation reactions already, they can continue even in the dark.

It may feel daunting to distinguish between PUR ester and ether foams and for many collections it will not even be possible because of limited or non-existent

Table 4.2 Possible Signs of Degradation for the Five Problem Plastics

Plastic	*Possible Signs of Degradation*
Hard rubber	Turns brownish/green, dulls. May form sulfuric acid so risk to metals and organics.
Cellulose nitrate	Cube-like fissures, yellowing, crazing, cracking, warping, weeping, whitish crystals, can become brittle. Releases acidic and/or camphor odor. May form nitric acid so risk to metals and organics.
Cellulose acetate	Weeping, cracking, warping, whitish crystals, can become brittle. May form acetic acid (vinegar smell) so risk to metals and organics.
PVC	Weeping, tacky, discolor, distort, whitish bloom, hardens. May form hydrochloric acid so risk to metals and organics.
Polyurethane	*Foam*: discolors, powdery, crumbles, stops bouncing back, may emit VOCs. *Fibers*: lose stretchiness, stops bouncing back.

access to scientific analysis. If that is the case, there are some general guidelines that apply to both types of PUR foams:

- Large surface areas increase vulnerability to degradation.
- PUR foams should NEVER be used for collection supports.
- Surface deterioration like discoloration, tackiness, or crumbling may be more obvious but the internal structure can also be degrading.
- Eventually, both types of foam lose ability to bounce back to shape.
- Sometimes foams can look fine but are actually already degrading. Since it can be difficult to tell when that begins and the process can be fast once it starts, avoid contact between foams and other materials or use interleaving products like silicone release Mylar/Melinex to prevent sticking.
- VOCs may be emitted so avoid keeping vulnerable materials like organics near PUR foams or counter the build-up of VOCs by increasing ventilation.
- PUR foams should be kept in the dark as much as possible but degradation can continue in the dark. If exhibited, display at low light levels.
- High RH environments should be avoided in case the foam is an ester.

Not All Problems are from Problem Plastics

It is important to remember that plastics not usually labeled as problems can also present collection care issues. Sometimes this is caused from manufacturing, exposures during use or from the storage environment, such as being exposed to off-gassing acids, or from other factors or a combination of causes. So, remember that plastics beyond the five problem plastics covered in this chapter can discolor, craze, weep, become tacky, release acids, crumble, and crack. Just because a plastic exhibits these signs does not mean they are a problem plastic and just because a plastic is not one of the problem plastics, does not mean it will age well and be benign to other materials.

Key Points for Problem Plastics

The deterioration mechanisms and storage/exhibition needs of the five problem plastics (HR, CN, CA, PVC, and PUR) are well understood. However, it can sometimes be difficult to identify a specific plastic type and then control the environment to promote its care, especially for collections that are not in large, well-funded institutions. But since these plastic risks are known, effort should be made to at least narrow down the possible identity of plastics, as discussed in Chapter 3, so targeted care decisions can be made, as covered in this chapter and further delved into in Chapters 5–7. Remember that changes to care may not require additional funding, such as reorganizing storage so that potentially problematic plastics are not stored with vulnerable collection materials or utilizing storage materials that may already be part of stocked supplies, such as buffered archival corrugated boards (blueboard). For some collections with problem plastics, however, decisions will have to be made as to whether or not they can remain or should be deaccessioned, as discussed in Chapter 9.

References

CAMEO, the Conservation and Art Materials Encyclopedia Online. 2022. "Materials Database." Museum of Fine Arts, Boston. https://cameo.mfa.org/wiki/Category:Materials_database

Derrick, Michele, Dusan Stulik, and Eugena Ordonez. 1993. "Deterioration or Cellulose Nitrate Sculptures Made by Gabo and Pevsner," In *Saving the Twentieth Century: The Conservation of Modern Materials*, edited by David W. Grattan, 169–182. Ottawa: Canadian Conservation Institute.

Farago, Robert. 2019. "SWATCH Big Bold Jellyfish." *The Truth about Watches.* December 29, 2019. https://thetruthaboutwatches.com/2019/12/swatch-big-bold-jellyfish/

Fenn, Julia. 1993. "Labeling Plastic Artefacts." In *Saving the Twentieth Century: The Conservation of Modern Materials proceedings of a conference Symposium '91 – Saving the Twentieth Century, Ottawa, Canada, 15 to 20 September 1991*, edited by David W. Grattan, 341–350. Ottawa: Canadian Conservation Institute.

Fenn, Julia. 2005. "Preservation of Plastics" workshop at Campbell Center for Historic Preservation Studies. August 15–18, 2005.

Halden, Rolf U. 2010. "Plastics and Health Risks." *Annual Review of Public Health* 31: 179–194. 10.1146/annurev.publhealth.012809.103714

Hornyák-Mester, Enikő, Dóra Mentes, László Farkas, Alpár Ferencz Hatvani-Nagy, Miklós Varga, Béla Viskolcz, Gábor Muránszky, and Béla Fiser. 2023. "Volatile Emissions of Flexible Polyurethane Foams as a Function of Time." *Polymer Degradation and Stability* 216. 10.1016/j.polymdegradstab.2023.110507

IPI (Image Permanence Institute). 2022. User's Guide for A-D Strips Film Base Deterioration Monitors. Rochester, NY: Rochester Institute of Technology. https://s3.cad.rit.edu/ipi-assets/publications/adstrips_instructions.pdf

IPI (Image Permanence Institute). 2024. "FilmCare.org, Storage Suitability." Image Permanence Institute, Rochester Institute of Technology. www.filmcare.org/storage_categories

King, Rose, Josep Grau-Bové, and Katherine Curran. 2020. "Plasticiser Loss in Heritage Collections: Its Prevalence, Cause, Effect, and Methods for Analysis. *Heritage Science* 8 (123). https://doi.org/10.1186/s40494-020-00466-0

Kodak. 2024. "Storage and Handling of Processed Nitrate Film." Kodak Motion Picture. www.kodak.com/en/motion/page/storage-and-handling-of-processed-nitrate-film/

Lattuati-Derieux, Agnès, Sylvie Thao-Heu, and Bertrand Lavédrine. 2011. "Assessment of the Degradation of Polyurethane Foams After Artificial and Natural Ageing by Using Pyrolysis-gas Chromatography/Mass Spectrometry and Headspace-solid Phase Microextraction-gas Chromatography/Mass Spectrometry." *Journal of Chromatography A* 1218 (28): 4498–4508. https://doi.org/10.1016/j.chroma.2011.05.013

Lauridsen, Clara Bratt, Lola Wøhlk Hansen, Theis Brock-Nannestad, Jesper Bendix, and Kim Pilkjær Simonsen. 2017. A Study of Stearyl Alcohol Bloom on Dan Hill PVC Dolls and the Influence of Temperature." *Studies in Conservation* 62 (8): 445–455. https://doi.org/10.1080/00393630.2016.1206651

Ledoux, Nicole, Georgina Rayner, Susan Costello, and Angela Chang. 2023. "Preventative Conservation, Treatment and Technical Study of Plasticized Poly(vinyl chloride) Multiples by Joseph Beuys." *Studies in Conservation* 68 (3): 343–356. https://doi.org/10.1080/00393630.2022.2033520

Metal Wonder Clarinets. 1897. "Metal Wonder Clarinets." Curatorial Files, Smithsonian's National Museum of American History.

MoDiP (Museum of Design in Plastics). 2007a. "Plastics Materials." Arts University Bournemouth. www.modip.ac.uk/projects/curators-guide/plastics-materials

MoDiP (Museum of Design in Plastics). 2007b. "Plastics Timeline." Arts University Bournemouth. www.modip.ac.uk/projects/curators-guide/plastics-timeline

Mulder, Karel and Marjolijn Knot. 2001. "PVC Plastic: A History of Systems Development and Entrenchment." *Technology in Society* 23 (2): 265–286. https://doi.org/10.1016/S0160-791X(01)00013-6

National Film and Sound Archive of Australia. n.d. "Storage of Cellulose Nitrate." National Film and Sound Archive of Australia. Accessed April 22, 2024. www.nfsa.gov.au/preservation/preservation-glossary/storage-cellulose-nitrate

Neves, Artur Louro Mendonça. 2023. "Cellulose Nitrate Objects in Collections: History of Science and Technology Hand in Hand with Conservation of Cultural Heritage." PhD diss., NOVA University Lisbon.

Plastic Identification Tool. n.d. "Plastics." Cultural Heritage Agency: Ministry of Education, Culture, and Science. Accessed April 19, 2024. https://plasticidentificationtool.nl/

Quye, Anita, David Littlejohn, Richard A. Pethrick, and Robert A. Stewart. 2011. "Investigation of Inherent Degradation in Cellulose Nitrate Museum Artefacts." *Polymer Degradation and Stability* 96: 1369–1376.

Royaux, Adeline, Emilande Apchain, Isabelle Fabre-Francke, Nathalie Balcar, Gilles Barabant, Clémentine Bollard, Bertrand Lavédrine, Odile Fichet, and Sophie Cantin. 2020. "Conservation of Plasticized PVC Artifacts in Museums: Influence of Wrapping Materials." *Journal of Cultural Heritage* 46: 131–139. https://doi.org/10.1016/j.culher.2020.07.002

Shashoua, Yvonne. 2003. "Effect of Indoor Climate on the Rate and Degradation Mechanism of Plasticized Poly(vinyl chloride)." *Polymer Degradation and Stability* 81 (1): 29–36. https://doi.org/10.1016/S0141-3910(03)00059-4

Shashoua, Yvonne. 2008. *Conservation of Plastics: Materials Science, Degradation and Preservation.* Oxford: Elsevier/Butterworth-Heinemann.

Shashoua, Yvonne. 2013. "Contemporary Conservation for Contemporary Materials." Paper presented at the American Institute for Conservation 41st Annual Meeting, Indianapolis, Indiana, June 2013.

Shashoua, Yvonne. 2014. "A Safe Place, Storage Strategies for Plastics." *Conservation Perspectives, the GCI Newsletter* 29 (1) Spring 2014. www.getty.edu/conservation/publications_resources/newsletters/pdf/v29n1.pdf

Townsend, Joyce H., Stephen Hackney, Mark Kearney. 2023. "A Critical Assessment of A-D Strips® Used in Degradation Studies of Artworks Made from Cellulose Acetate." *Studies in Conservation*, 68 (1): 9–17. https://doi.org/10.1080/00393630.2021.1996091

van Oosten, Thea. 2005. "Preventive Conservation." In *Course Notes from ICN Masterclass Working with Plastics*, 53–54. Amsterdam: Netherlands Institute for Cultural Heritage.

van Oosten, Thea. 2011. *PUR Facts Conservation of Polyurethane Foam in Art and Design.* Amsterdam: Amsterdam University Press.

van Oosten, Thea. 2022. *Properties of Plastics: A Guide for Conservators.* Los Angeles: Getty Conservation Institute.

Williams, R. Scott. 1994. "Display and Storage of Museum Objects Containing Cellulose Nitrate." *Canadian Conservation Institute (CCI) Notes* 15/3. www.canada.ca/en/conservation-institute/services/conservation-preservation-publications/canadian-conservation-institute-notes/display-storage-objects-cellulose-nitrate.html

Williams, R. Scott. 2002. "Care of Plastics: Malignant Plastics." *WAAC Newsletter* 24 (1) January 2002. https://cool.culturalheritage.org/waac/wn/wn24/wn24-1/wn24-102.html

Williams, R. Scott. 2019. "Plastic Storage Products." In *Preventive Conservation: Collection Storage*, edited by Lisa Elkin and Christopher A. Norris, 753–799. New York, NY: Society for the Preservation of Natural History Collections.

5 Influencers on Plastics

Environment

This chapter delves into the main environmental issues that can affect plastic collections: light [visible, ultraviolet (UV), and infrared (IR)], temperature, relative humidity (RH), and pollutants. It is important to understand options for monitoring and mitigating these environmental factors since they can have a significant impact on the preservation of plastics.

Light

We need light to see, but we do not need all spectra of light to see, only light in the visible wavelength. But since the edges of the visible spectrum blur into IR and UV, as shown in Table 5.1, it can be difficult to fully eliminate them. Some light sources emit IR or UV light, which introduces risks to some materials like plastics. It is important to remember that though necessary, visible light can also damage. It is imperative to evaluate the type of light, know how much of it is reaching vulnerable plastics, and determine options for reduction or elimination.

Key factors to keep in mind with light exposure are that light damage is cumulative and irreversible. Photochemical reactions, including photo-oxidation, the chemical breakdown caused by light and oxygen, are drivers in the degradation of several types of plastics (Table 5.2), but all plastics should be kept in the dark when not exhibited. Light damage typically manifests visually as discoloration: yellowing, darkening, or fading. But the damage is not only at the surface; photochemical reactions accelerate degradation of the polymer itself, making it weaker.

Light Sources

Look around an exhibition. What are the sources of light? Is there natural sunlight entering through windows or skylights? Are there overhead lights giving a somewhat even lighting effect? Perhaps there are targeted spotlights protruding from the ceiling or along the upper edge of a wall. Are there lights within display cases? Is there a combination of these? Knowing where light is coming from and the type of light is vital for determining risks to plastics.

DOI: 10.4324/9781003489658-5

Table 5.1 Types of Light

Type of Light	*Wavelength (nm)*
Ultraviolet	100–400
Visible	400–750
Infrared	750–2500

Table 5.2 Degree of Plastic Sensitivity to Light

Plastic	*Visible Light*	*Ultraviolet Light*
Acrylonitrile butadiene styrene (ABS)	High	High
Cellulose acetate (CA)	Medium	Medium
Cellophane (CE)	Medium	Medium
Casein formaldehyde (CF)	Medium	Medium
Cellulose nitrate (CN)	Medium	Medium
Epoxy (EP)	High	High
Ethylene vinyl acetate (EVA)	Medium	Medium
Natural rubber (NR)	High	High
Polyamide (PA)	Medium	Medium
Polybutylene adipate terephthalate (PBAT)	Medium	Medium
Polyethylene (PE)	Medium	High
Polypropylene (PP)	Medium	High
Polystyrene (PS)	Medium	High
Polystyrene (PS) foams	Medium	Medium
Polyurethane ester (PUR ester)	Medium	Medium
Polyurethane ether (PUR ether)	High	High
Polyurethane expanded (PUR expanded)	Medium	Medium
Polyurethane flexible (fibers) (PUR flexible)	Medium	Medium
Polyvinyl chloride, plasticized (PVC-P)	High	High
Polyvinyl Chloride, unplasticized (PVC-U)	Medium	Medium
Styrene acrylonitrile (SAN)	Medium	Medium
Synthetic rubber (SR)	High	High
Unsaturated polyester (UP)	High	High

Source: Plastic Identification Tool (n.d.).

Sunlight entering through windows, a skylight, or glass entry doors contains both visible and UV light. UV light is found on the blue/violet end of visible light and has shorter, more energetic wavelengths. If you picture these shorter wavelengths as quickly pounding with more energy as opposed to being slower moving, it becomes evident why UV light is damaging. UV light is not needed to see, so it should be filtered out. In addition to sunlight, fluorescent lights, often in the form of long tubes like those commonly found in office ceilings, can be an indoor source of UV light, though the levels emitted vary depending on the bulbs. Since fluorescent lights are being phased out in many areas, the risks their presence brings to collections will diminish over time.

On the other end of the visible spectrum, IR light begins. Sources of IR light, namely incandescent and tungsten halogen bulbs, are becoming obsolete because they are not energy efficient. They are better producers of heat than light, which can be a collection concern, especially in an enclosed space like a case. In addition to IR, these bulbs can also emit UV light, which compounds concerns with their use (Himmelstein, Rosenfeld, Weintraub 2019).

The preferred source of visible light in museums is light-emitting diodes (LEDs) because they use less energy to run, so they support sustainability goals; last a long time, so they are more economical; come in a range of colors to get different lighting effects; and the beam spread can be broad or targeted. LEDs do not emit IR but do create some heat, though it is markedly less than other light systems. Another advantage of LEDs is that they do not emit UV light, but this should be confirmed by the manufacturer before using for collections (Himmelstein, Rosenfeld, Weintraub 2019).

Light Monitors

Since it is important to know the type of light present and the intensity of it, light levels should be monitored and tracked over time. The ability to do this ranges from handheld electronic monitors that provide spot readings to durational monitors that can continuously record light levels uploaded to a central computer-based platform. An advantage of electronic light monitors is that they provide data. Ideally, data is collected and interpreted to determine how much light exposure collections are receiving, how that changes with the seasons, and what levels they are exposed to if loaned to another institution. And then this data should be used to track cumulative light exposure over time, as discussed in the mitigation section below.

Visible light levels are most commonly measured in lux which is the International System of Units (SI) value that equates to 1 lumen per square meter. Some museums, particularly in the United States, may use the imperial system measurement, foot-candle, which is one lumen per square foot. If you have to convert between the two units, 1 foot-candle = 10.76 lux.

Though ideally filtered out, UV light may be present and, therefore, should be monitored. The SI unit for measuring UV light is microwatts per lumen, and the imperial system measurement is milliwatts per square meter. For plastics, if UV light is present, it should be below 30 microwatts per lumen (MoDiP 2022).

Light monitors range in cost and features, with some units able to monitor both visible and UV light, while some only detect within one range. If purchasing a light monitor for either visible or UV light or both, it is best to buy from a trusted supplier who works with museum collections.

For museums with few collection staff, monitoring light levels with the International Standards Organization (ISO) Blue Wool Standard card may be a good option because the cards do not require much staff time for deployment or interpretation. The Blue Wool Standard consists of eight strips of blue wool cloth attached to a card. Each wool strip is dyed with blue dyes of different lightfastness that fade at known rates, with each strip needing twice as much light exposure as the one before it to fade to the same degree (CAMEO 2022). Because plastics have the potential to be very light sensitive, they should be considered equivalent to ISO Blue Wool Standards # 1, 2, and 3 (Himmelstein, Rosenfeld, Weintraub 2019). So, when those most sensitive blue wool strips start to fade, you can assume the maximum recommended light exposure for a plastic has been reached under the current conditions. It is important to remember that if the more energetic UV light is present, the Blue Wool dyes will fade quicker. If you are curious how effective UV film is at slowing the fade rate of Blue Wool Standards, you could cut a strip of UV-blocking film and cover part of the Blue Wool card. The results will show that while UV filters slow the rate of fading somewhat, there will still be a color change in the most sensitive dyes because of exposure to visible light.

If your budget is truly limited and Blue Wool Standard cards cannot be purchased, a quick, inexpensive way to get a general sense of potential light damage is using readily available office supplies that have fugitive dyes, namely brightly colored fluorescent highlighters (yellow and pink are good choices) and/or sticky notes (bright pink usually works well). Often the yellow highlighter will darken to brown, and the pink fluorescent dyes in the highlighter and sticky notes fade. Because the dyes used are so fugitive, they are sensitive to light and quick to change. If the monitors change color, it is a pretty safe assumption that the light levels are too high for sensitive plastics. The quicker the change, the more intense the light, which is likely accelerating photochemical reactions.

To set up this do-it-yourself light monitor:

- Use a thick piece of white paper, cardstock, or index card
- Cut to desired size
- Draw horizontal lines with the highlighter(s) and/or adhere the sticky note(s)
- Cover half of the card with aluminum foil, leaving half of the color samples exposed. *This enables direct comparison to the exposed areas to monitor for change.*
- Place the card in the location that you want to know more about potential light damage
- Monitor at least once a week for several months to compare changes from the original color

Color Monitors

If available, a spectrophotometer can be used to calculate color on the plastic itself and note any changes over time using the color system developed by the International Commission on Illumination (CIE) known as CIELab which uses three values, L*a*b*:

- L* plots lightness on a scale of 0 (black) to 100 (white)
- a* refers to red–green values; negative values are greener, positive ones are redder

- b* values plot the blue–green range; negative values are bluer, positive ones are yellower
- ΔE combines all three values

If color data is collected at the time of accessioning, before an exhibition, or really anytime, and then periodically monitored, the subjectivity of determining color change of the plastic is removed because the color is plotted on the L*a*b* scale, and then the ΔE value can be used to quantify and compare change, as discussed in Sidebar 8.3. Though not every degradation mechanism will cause color change, those vulnerable to photo-oxidation likely will change over time, and recording this alteration will help flag and track damage and, ideally, help guide preventive care.

Light Mitigation

As seen in Table 5.2, because some plastics are severely damaged by light, it is a good idea to protect them from unnecessary light exposure in storage as well as when exhibited. Generally, it is recommended to display plastics at about 50 lux or 5 foot-candles (that equation is not accurate given the conversion stated above, but typically these are the levels given for light-sensitive materials for ease of setting lighting).

Keep in mind that damage sustained by an object is the same from a long exposure at low light levels as it is from a short exposure at high light levels, which is referred to as the principle of reciprocity (Figure 5.1). So, displaying at 50 lux for 200 hours is the same as 200 lux for 50 hours, both totaling 10,000 lux hours. Note that 200 lux is typically the upper limit for display lighting and is reserved for non-light-sensitive materials like metal and stone, meaning that plastics should never be exhibited under such bright conditions.

Since light damage is cumulative and irreversible, it is important to keep track of the total light exposure for an object over time. For example, exhibiting at 50 lux for 10 hours a day for 365 days a year equates to 180,000 lux hours. At some museums, light exposure is tracked in the collection database to ensure that light-sensitive materials that are frequently requested for exhibition get periodic "rest" from light.

Depending on the space and light source, light reduction could come in the form of sensors so lights are only on in a room when movement is detected or by installing timers so if lights are on after a specified time period, such as after the

Figure 5.1 Principle of reciprocity for light exposure.

museum is closed, they will automatically turn off. Since light reduction methods like sensors and timers decrease energy use, support for their installation may be sought from energy-savings or sustainability initiatives.

In some principalities, there will be laws requiring safety lighting. If the storage room or exhibition spaces used for your collections have safety lights that remain on 24 hours a day, determine if that light is reaching collections. If it is, can you move vulnerable objects? If it is in storage and the items are on open shelving or in glass-front cabinets, can the light be physically blocked with the addition of a stable, opaque material like heavy cardstock in front of the object?

If sunlight comes through, are there UV-blocking films on the windows, doors, or skylight? UV film will become exhausted over time, so it is a good idea to periodically take readings at the source as well as at the collections that the light hits to determine if the UV films are still effective or need to be replaced. If fluorescent lights are a source of UV, UV-blocking sleeves that go over the bulbs or UV-filtering lenses on the fixtures themselves may be a mitigation option. Before installing UV filters, though, make sure they block wavelengths below 400 nm since this is not the industry default (Himmelstein, Rosenfeld, and Weintraub 2019).

Temperature

High temperatures drive chemical reactions, with generally the rate of chemical reactions doubling for every 7.8°C or 18°F rise in temperature (Simione 2019). So, storing plastics at lower temperatures will slow the rate of chemical deterioration. Some plastics families are particularly vulnerable to thermal oxidation even at room temperatures (20°C or 68°F) (Williams 2019). These include:

- Polyethylene (PE)
- Polypropylene (PP)
- Polyvinyl chloride (PVC)
- Polyurethane (PUR) ether foam
- Cellulose nitrate (CN)

Remember that the glass transition temperature (Tg) is the temperature when a rigid plastic softens, which is another factor for some plastics at higher temperatures (Chapter 3). For plastics like CN film, frozen or cold storage will slow chemical reactions, keep it below its Tg, and prevent the ignition point from being reached (Chapter 4).

As temperature increases, air can hold more moisture, which lowers the RH. As temperatures drop, the air has less energy to hold moisture, so the RH increases. So, essentially, temperature and RH are inverse. In temperate climates, this phenomenon will be familiar indoors as hot, dry attics and cool, damp basements. Even in a well-sealed museum case, this inverse relationship of temperature and RH will hold because the environment within the case will eventually react to changes in the surrounding room. Generally, an increase in room temperature of

1°C (1.8°F) will lower the RH by about 3%, and a decrease in room temperature by 1°C (1.8°F) will result in about a 3% increase in RH levels (Weintraub 2019).

Temperature Sources

Generally, museums aim to keep their indoor spaces in the range of 15–25°C (59–77°F), but many institutions still aim for the once more common recommendation of about 20–22°C (68–72°F), which is driven by human comfort rather than collection preservation. The Museum of Design in Plastics (MoDiP) recommends that synthetic garments be stored below 20°C (68°F), which is a good guideline for plastics in general (2022).

Changes in indoor temperatures outside the bounds of museum goals are typically the result of malfunctioning heating, ventilation, and air conditioning (HVAC) units or the discrete units that heat or cool, like window air conditioners. For institutions that do not have mechanical units to regulate temperature, changes will follow the seasons. Indoor causes of increasing temperatures may come from light sources (see the Light section) as well as from times of increased visitor numbers.

Temperature Monitors

Since elevated temperatures can be damaging, environmental monitoring in locations that house plastics is important, whether that is within a case or at the room level. If the museum has an HVAC, it can be tempting to assume that temperature is already known because of the thermostat, which is typically wall-mounted. If the thermostat is only reading that one area and if its readings are controlling multiple rooms, actual conditions can vary. To better understand the environmental conditions in the spaces that house collections, more monitoring is likely needed.

Temperature monitors are frequently coupled with RH and come in options for either spot measurements or continuous monitoring. Spot measurements can be taken by handheld devices, with some units also having RH as well as visible and UV light levels able to be detected. Analog monitors, such as dial-based thermohygrometers or digital options that have electronic displays, are available at a variety of price points. With these devices, someone has to regularly record and interpret data for them to be useful.

The once common hygrothermographs, sometimes called thermohygrographs, which record temperature and RH with ink on paper wound on a cylinder, have the advantage of continuously charting data, though the paper needs changing and the ink eventually runs out. But as with spot measurements, someone has to interpret the data.

Data loggers that automatically record data are becoming more common in collection care. The data collected by these monitors is taken at intervals and is then either manually uploaded or the sensors may use Wi-Fi or cloud storage that can be accessed through an online interface. It is usually easy to generate temperature and RH graphs, typically with the user able to hone in on varying periods of time such as a day, week, month, season, year, or the full duration of time the sensors have been deployed. A key advantage of electronic data loggers is that alerts can be sent if the temperature or RH goes outside a set range. Of course, sending alerts is only useful if they go to people who will see them and are empowered to act.

If cool, cold, or frozen storage is being used (more in Temperature Mitigation), monitors that alert if the temperatures go out of bounds and people who can address the issue are vitally important for the success of low-temperature storage systems.

Temperature Mitigation

The building is the first line of defense against changes in the environment. If the building envelope is not well-sealed, exterior changes over the seasons or from inclement weather will affect the indoor environment, which, in turn, affects collections, as well as the people who work and visit. Auditing the building envelope is a good place to start to mitigate temperature issues. Can gaps in windows and doors be better sealed? Can insulation be added? Can light-blocking curtains or scrims be added to windows to reduce heat load as well as lower light levels?

Once the building envelope is improved, is it necessary or even possible to lower the temperature of rooms that house collections (this may not be possible if the HVAC is zoned and mixed-use spaces are controlled together)? It is important to determine if decreasing the temperature will increase the RH to unacceptable levels. For example, you would not want to lower the temperature to the point of condensation.

If whole room temperature solutions are not possible, can certain vulnerable plastics be targeted for cool, cold, or frozen storage (see Chapter 4, Table 5.3, and Sidebar 5.1)? Creating and maintaining lower temperature storage requires funds, has limited space, and uses more energy, so its utilization should be made in consultation with a variety of museum stakeholders. For example, the collection care team may flag vulnerable plastics, especially ones that have inherent hazards like CN film, while the curatorial staff may advocate for plastics that are central to the museum's mission to go into lower temperature storage in order to preserve them for a longer time, and the administration may balk at the request because of the cost or because it is counter to energy-reduction goals at the museum. It is important to note that while low-temperature storage has its benefits, it comes with risks for some plastics. As examples, polycarbonate becomes brittle at low temperatures (Plastic Identification Tool n.d.) and a study from The Costume

Table 5.3 Temperatures Ranges for Low-Temperature Storage

Low-Temperature Term	*°C Ranges*	*°F Ranges*
Cool	12.8–18.3	55–65
Cold	0.56–12.8	33–55
Frozen	0 to −40	32 to −104

Source: Simione (2019).

Institute found that silicone, cellophane, and 3D-printed resin had greater changes in color at low temperatures than in standard museum storage conditions (Garcia, Hays, and Scaturro 2020). The same study also found that plasticized PVC, neoprene (chloroprene), and nylon (polyamide (PA)) had more color change in frozen storage than in cold storage (4.4°C or 40°F), but polyester, PUR ether foam, and PUR elastomer fared better at frozen temperatures (Garcia, Hays, and Scaturro 2020). So, as with so much in collection care, decisions need to be discussed and pros and cons determined before action is taken.

If cool, cold, or frozen storage is used, there are practical measures to keep in mind. Make sure to have sensor alerts for both temperature and RH; the Image Permanence Institute (IPI) assumes RH in the range of 30–50% for all storage, including low-temperature ranges (2024). With low-temperature storage, it is essential to have a back-up generator in case of power loss. Also, determine protocols for re-entering these plastics into the ambient environment. Possible options could be double bagging, typically with low-density polyethylene (LDPE) sealable bags, so condensation forms on the outside bag when brought in or out of room temperature, having a cool intermediate space to help with acclimation, or using insulating material like polystyrene (PS) chips to act as a buffer against stark changes (Shashoua 2014). The critical moisture indicator (CMI) packaging method accounts for the potential for a thermal difference in the packaging to cause hygroscopic materials, such as negatives or photographs, within to release or absorb moisture. To counter this, the object is not only double bagged; it is also stored with a "moisture trap" like mat board between the bags to buffer the RH and monitored with humidity indicators, such as those discussed later in this chapter (McCormick-Goodhart 2003). Remember, if low-temperature storage is utilized, the need to acclimate back to the ambient environment means that access to that collection item has to be planned for ahead of time.

Sidebar 5.1 The Positive Results of Rehousing a Flexible Polyvinyl Chloride Object

Susan Costello, Nicole Ledoux, and Georgina Rayner

The Harvard Art Museums began rehousing its flexible PVC objects after encountering three Joseph Beuys multiples in advanced states of degradation. By 2012, all three of these artworks were covered with droplets and pools of liquid plasticizer, and documentation over the course of the preceding decade revealed the progressive worsening of their condition. Through research, technical study, and experimental testing, preventive conservation solutions were developed and implemented for these objects (Ledoux et al 2023). In 2016, the museum embarked on the first comprehensive survey of plastic works of art in its collection. During the survey, a conservator assessed the condition of the work of art, and a scientist identified the plastic type using analytical instruments when sampling was possible. The survey data revealed that PVC is the most common plastic in the collection. PVC is one of the most at-risk plastic types in museum collections because its unstable chemical composition causes it to deteriorate rapidly. As a result, this group of objects was prioritized for intervention. For this material, preventive measures, including rehousing and improving storage conditions, are the principal ways to extend the life of the object.

Different plastic types require carefully chosen housing materials that serve a specific function. For flexible PVC, the optimal housing material and design should prevent the loss of plasticizer from the material. To make the plastic pliable, flexible PVC contains up to 50% plasticizers, which are not chemically bonded to the PVC structure. This manufacturing technique leaves the PVC inherently unstable. As the plastic ages and becomes more crystalline, there is less space for the plasticizer. It migrates to the surface, forming pools of hazardous liquid that slowly evaporate. The process continues until the plastic loses so much plasticizer that it becomes brittle and cracks. Studies have shown that PVC benefits from low-temperature storage

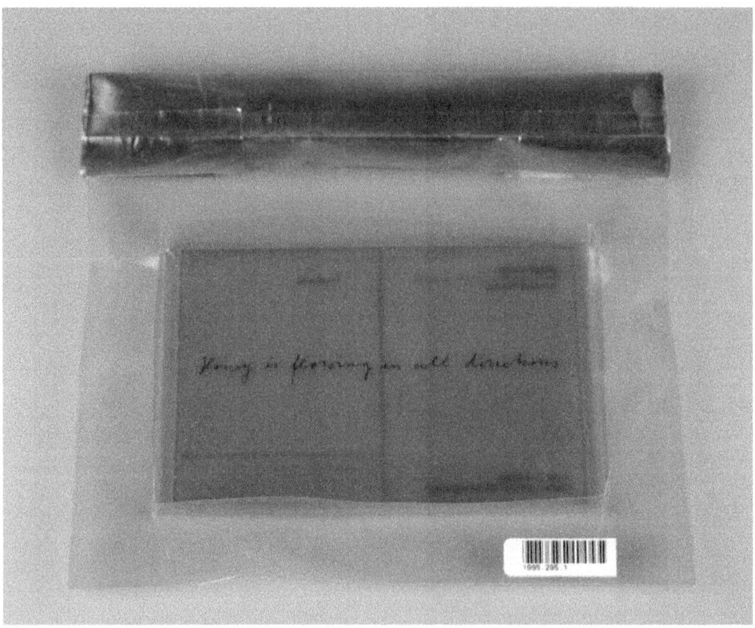

Figure 5.2 Joseph Beuys, *Postcards 1968–1974*, 1974, wood, PVC, and cardboard, 18 × 12.5 × 6 cm^3. Harvard Art Museums/Busch-Reisinger Museum, 1995.295. PVC component rehoused in Mylar to prevent the loss of plasticizer. Image courtesy of Harvard Art Museums.

in non-absorbent materials like glass or polyester film (Shashoua 2008; Ledoux et al 2023). Reducing the storage temperature slows the rate of plasticizer migration to the surface, while a non-absorbent enclosure allows plasticizer vapor to accumulate until the internal environment is saturated, preventing further evaporation.

Positive results have been observed for one of the PVC objects by Joseph Beuys (1921–1986) entitled Postcards 1968–1974 (Ledoux et al 2023). In 2014, the PVC component was rehoused in a Mylar sleeve sealed with aluminum foil tape (Figure 5.2). The rehoused object was placed in cool storage (55°F, 40% RH). When examined five years later, most of the plasticizer appeared to have been reabsorbed by the PVC. This observation suggests that the polymer structure has not become completely crystalline, leaving space for plasticizer to reintegrate (Yvonne Shashoua, Research Professor in Synthetic Polymeric Materials, National Museum of Denmark. Personal Communication. December 9, 2019). However, since this result has not yet been observed for other similar objects, further study will be required to better understand this phenomenon.

Upgrading the storage of this object has dramatically improved its condition. The degradation of PVC cannot be stopped, but the successful rehousing of this object demonstrates that we can extend the lives of these unstable plastics.

In exhibition galleries, room temperature is going to be governed by human comfort and what your building can achieve. In some areas of the world, there is a greater expectation that a museum be similar to the temperature of a comfortable living room, while in other places, visitors will be fine wearing a jacket in galleries on a cold winter day. For temperature-sensitive plastics on display, it may be possible to design a microclimate case that more tightly controls temperature and RH (discussed below). However, if that is the route taken, make sure that off-gassing from plastics or the case materials themselves do not build-up in the enclosed space (see Pollutants section).

Lowering temperatures indoors, whether at the building, room, or case level requires a lot of energy. The rising temperatures in many places as the result of the climate crisis mean that advocating for high-energy-consuming methods such as HVAC or dedicated low-temperature storage spaces may not only be hard to get support for; it may be deemed irresponsible or even unethical. As with so much in collection care, it is therefore essential to incorporate various viewpoints so that a fully informed decision can be made for temperature control.

Relative Humidity

RH is the amount of water vapor in air compared to the maximum amount the air could hold. It can be helpful to think of RH as relative saturation. Because this amount is relational or proportional, RH is expressed as a percentage. Higher temperatures retain more moisture, so RH decreases, and low temperatures hold less moisture, so RH increases.

Guidelines for general museum environments for mixed collections typically aim for a set point in the 45–55% RH range with an allowable drift of +/−5%, which works out to an annual RH range of 40–60%. These general recommendations are suitable for many plastics, but there are exceptions.

As discussed in Chapter 4, the upper part of this general RH range is harmful to plastics vulnerable to hydrolysis, a chemical reaction with water that causes the polymer to break down back to its monomer (Williams 2019). As this reaction occurs, the plastic can craze, warp, and crack. Plastics vulnerable to moisture-driven degradation should be kept below 50% RH, ideally within a stable setpoint within these ranges (Plastics Identification Tool, n.d.):

- Cellulose acetate (CA), 20–30% RH
- Cellulose nitrate, 20–30% RH
- Polyurethane esters, 10–30% RH

For some plastics, RH that is too low can cause issues. MoDiP recommends that the RH for synthetic fibers stay above 40%, noting that cellulose triacetate (CTA), nylon, polyester, and acrylic are vulnerable to creating static charge at low RH (MoDiP 2022). At low RH, casein formaldehyde (CF) and urea formaldehyde (UF) can desiccate and shrink and are best kept at 60% RH, +/− 5% (Plastic Identification Tool n.d.).

RH fluctuations can damage plastics with RH sensitivities, causing strain from taking on and losing water, resulting in crazing and warping (van Oosten 2022). Therefore, not only is it important to have plastics in their preferred RH ranges but it is best if fluctuations are minimal. Even plastics that are considered to be stable like polymethyl methacrylate (PMMA), which is likely known to museum staff by names like acrylic, Plexiglas or Perspex, can be damaged by RH fluctuations. So, rather than requiring a higher or lower RH setpoint, it is best kept at whatever RH can be most consistently maintained within general museum standards to avoid damaging fluctuations (Plastic Identification Tool n.d.). If a plastic is degrading and additives are weeping out, fluctuating RH can cause changes to those leached products, which can stress the plastic. For example, low RH can cause these products to crystalize within the plastic which may put physical stress on the plastic as the crystals take up more room. The exuded additives in Figure 5.3 were liquid at higher RH (Figure 4.2) and crystalized when moved to an environment with lower RH.

Relative Humidity Sources

As with temperature, RH fluctuations are often the result of seasonal changes or weather. The degree to which those are mitigated inside depends on the building envelope and the location of the room within the building. Rooms with external walls, especially those with more windows, have increased vulnerability to

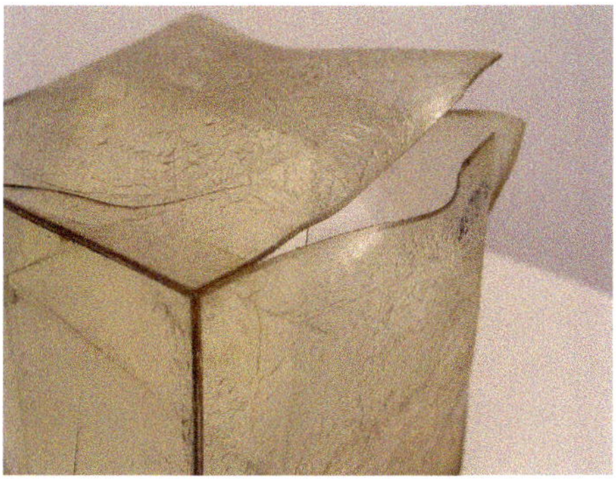

Figure 5.3 Circa 1930s geometric model made of a CN internal form and CA external form. The leached plasticizers from the CA crystallized at lower RH. Once in the NMAH mathematics collections (1979.0102.058). Photographed by author, Division of Medicine and Science, National Museum of American History, Smithsonian Institution.

mimicking outdoor conditions if not well insulated. Unless specifically equipped to regulate humidity, most HVAC units are designed for human comfort, which is dictated by temperature, not RH. This means that a system that aims to meet the previously stated goal for plastics collections of not exceeding 20°C (68°F) could result in the RH plummeting to single digits in the winter as the heat comes on and rising beyond 70% in summer as the air is cooled. While people may still feel comfortable, those large RH fluctuations can damage collections, with some plastics being especially vulnerable.

Relative Humidity Monitors

Options for analog, digital, and wireless monitoring for RH are the same as already discussed for temperature. For RH monitoring, an additional inexpensive option exists: humidity indicator cards. These come in a standard option that reads the current RH with the viewer determining the RH at 10% increments with instructions like "read top of blue between blue and pink" or "read at lavender between blue & pink" with blue being drier, lower RH, and pink indicating higher humidity (Talas 2024 and Gaylord Archival 2024). These standard cards are sometimes referred to as reversible because they can be continually used. It takes about 24 hours for them to react to RH changes, but the fact that they can be continually used is appealing as is their relative degree of accuracy of ±2.5% RH at 75°F (24°C) (Cole-Parmer n.d.). These cards are likely best purchased from conservation and collection care suppliers, but they are available from other sources. Before ordering, make sure that the cards read in the higher RH ranges since that is typically the concern for plastics; some humidity cards are focused in the 10–60% range, which is too limited for plastics.

Another type of humidity indicator known as Maximum Humidity Cards detect the highest humidity level reached in the 50–90% RH range. These work by:

> Each level of humidity is represented by a blue crystal that dissolves to create a large blue spot that clearly indicates the highest point of relative humidity that has been reached for a period of time greater than 24 hours.
>
> (Talas 2024)

If the RH reaches 50% or higher for 24 hours or more and the crystal dissolves, the card is used up since the reaction is irreversible. While this may seem limiting, it can still be an important tool because the cards are likely not continuously monitored, and knowing if the 50% threshold was reached can be important for some vulnerable plastics. Some suppliers offer the maximum, irreversible humidity cards coupled with the standard, reversible humidity indicators.

Relative Humidity Mitigation

Since RH and temperature are interconnected, approaches to mitigation are similar, namely assessing the building envelope and ensuring that any mechanical systems that regulate temperature and RH are in good working order with filters that are regularly changed.

A layered approach such as an object in a storage tray, within a sealed cabinet, within a room (ideally interior) that is within a building with environmental controls will slow RH fluctuations that reach the object. Within cases, hygroscopic materials, which could be organic collection items like wood and paper or housing materials like blueboard, can absorb or desorb moisture in response to changes in RH in the surrounding environment. Since these materials want to be at equilibrium with the environment, they can help to self-regulate moisture in a case (Weintraub 2019) (Sidebar 5.2). The rate of equilibrium moisture content (EMC) can take days or weeks to reach, meaning that RH changes are slower to take effect than thermal equilibrium, which is usually reached in only hours. As a reminder though, temperature and RH impact each other because of their inverse relationship. Because cases are microclimates, they need to be monitored if sensitive plastics are within.

If an environment is too humid to safely store sensitive plastics, it may be necessary to use dehumidifiers even if the objects are within cases. Some HVAC units are capable of removing moisture from the air to reach a set point, but it may not be fully effective or struggle during some seasons. If a targeted room-level approach is needed, portable dehumidifiers may be an option. If these are utilized, ensure that they are either routinely emptied, which in some climates may be several times a day, or, ideally, set them up to automatically drain. Since large amounts of water can be pulled out in dehumidification, water alerts should be set up around the unit to alert if they leak. If dehumidification is not possible, the addition of fans to circulate air can be helpful.

For conditions that are too dry for some plastics, it may be necessary to humidify either through the HVAC system or at the room level. Portable units require the addition of water, so staff time is needed for this task, and there is a risk of leaks. Placement needs to be determined to avoid potential condensation. Remember that lower temperatures hold less moisture, so if the humidified air hits a cold surface, it can condense out, meaning that a portable humidifier near an exterior wall would be a poor placement choice in cold climates.

At the case level, preconditioned silica gel may be an option for slowing RH changes, but its effectiveness will depend on how well the case is sealed. If the case is leaky, the silica gel will be rapidly exhausted and, therefore, would be an infective solution (Weintraub 2019). An additional factor to consider for plastics in cases is that if the plastics or the case materials are releasing acidic vapor or volatile organic compounds (VOCs), then a tightly sealed case that traps these degradation products and can catalyze reactions could be harmful. So, before employing silica gel, it is a good idea to check for off-gassing, as discussed in Chapter 7, and measure case leakage. Testing case leakage is still not standard practice in most museums but its utilization is increasing. Essentially, gas leak detectors like those used by HVAC technicians, can be employed to determine where leaks are in case seals so that they can be fixed (Weintraub 2019).

If a case is leaky and either cannot be fixed or if it is preferable to improving the seals and permitting off-gassed products to build up, electro-mechanical equipment options exist to actively control RH (Weintraub 2019). Because this is specialized equipment, it is a more expensive option but may be a good solution in some situations.

Sidebar 5.2 Physical Response of Plastics to the Environment: Equilibrium Moisture Content (EMC)

Emma J. Richardson and Jacek Olender

While the field has developed methodologies for the identification and characterization of plastics, an in-depth understanding of chemical and physical degradation and possible mitigation strategies for collection care are currently in their infancy. What is well understood is the dependence of the physical properties of many plastics on both temperature and moisture. Environmental fluctuations cause physical change to many plastic objects owing to changes in moisture content as they equilibrate to their surroundings. The resulting expansion and contraction do not necessarily result in damage, with the impact to the object often reversible. However, under certain circumstances, irreversible physical change occurs, often displayed as buckling, crazing, and delamination. Therefore, the management of storage, display, and transit environments appropriate for plastics can serve to reduce the risk of physical degradation.

To understand the response of plastics to their environment, it is important to define a few key terms. The equilibrium moisture content (EMC) represents the amount of moisture in hygroscopic materials when they are at equilibrium with their ambient environmental conditions. Changes in temperature and RH cause hygroscopic materials, such as plastics, to absorb and desorb moisture accordingly. The RH is the ratio of water carried by air relative to the *total* amount of water it can carry at the same temperature. Similarly, the equilibrium relative humidity (ERH) is the amount of water carried by air surrounding a material when in *equilibrium* with the water in the material. At equilibrium, the ERH is directly proportional to the RH and moisture content.

If RH and ERH are equal to each other, then the moisture content of a plastic object will remain constant. However, if the RH of a collection or storage space is lower than the localized equilibrium value, then water will move from the object to the air, reducing the plastic's moisture content and increasing the RH of the air. The opposite is true if the RH of the space is higher than the ERH. Changes in temperature will also influence the moisture content of plastics, such that at lower temperatures, water will be more tightly held by the plastic, and vice versa. Therefore, to maintain a constant

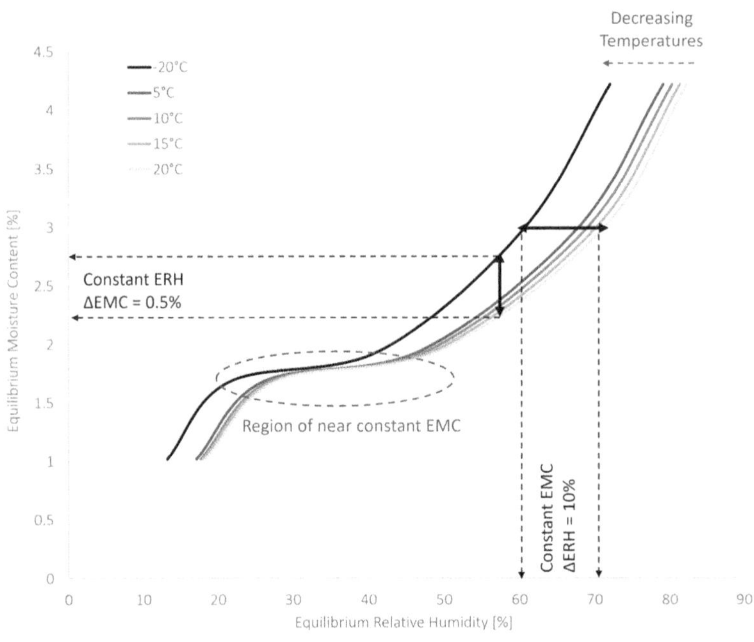

Figure 5.4 Isotherm demonstrating the relationship among temperature, equilibrium relative humidity, and equilibrium moisture content for a cellulose triacetate film plasticized with 8% diethyl phthalate and 8% triphenyl phosphate (wt%).

moisture content when changing storage and display temperatures, the surrounding RH needs to be altered.

Knowledge of the relationship between a plastic's EMC and ERH at different temperatures can be helpful to understand safe storage conditions for plastics objects. Figure 5.4 illustrates this relationship for a CTA film containing two different plasticizers, namely diethyl phthalate and triphenyl phosphate. These isotherms, referring to the interconnecting lines of constant temperature, demonstrate that there are temperature and ERH combinations within which changes in moisture content remain constant or are greatly reduced. These graphs can be used to assess the impact of changing environments. For example, if moving a triacetate-based object between 20 and −20°C, the moisture content can be maintained near constant if the ERH is between 25 and 45%. Within a collection space, the environment will be dynamically changing, but at the object level, ERH will be equivalent to RH. Similar isotherms are available in the literature for other classes of materials, which can be used to inform collections decisions.

It is important to reiterate that not all expansion and contraction in plastic objects is problematic. Many objects can withstand changes in their dimensions without necessarily causing permanent damage. Therefore, understanding what degree of fluctuation in moisture content can be physically tolerated can aid sustainable collections management of plastics. Quantifying change for plastic composite objects and understanding the tipping points between safe and damaging environmental transitions will support collecting institutions in implementing less-restrictive, energy-saving strategies that drive down costs without compromising on collection care. Please see Sidebar 5.3 on hygroscopic expansion coefficients for further details.

Sidebar 5.3 Physical Response of Plastics to the Environment: Humidity Expansion Coefficient (HEC)

Jacek Olender and Emma J. Richardson

It is a well-known phenomenon that hygroscopic materials expand and contract as they absorb and lose moisture. The majority of organic materials housed in collections, including plastics, are hygroscopic in nature and will physically respond to changes in atmospheric moisture. However, current understanding of the impact of the repeated expansion and contraction of plastics due to changes in surrounding RH is still limited. Dimensional changes are often small, although they have the potential to be damaging if an object is physically restrained or if component materials have different dimensional responses to each other, such as laminate structures and painted surfaces.

As discussed in Sidebar 5.2 on EMC, organic-based materials such as CA objects do not absorb water from the atmosphere linearly. Therefore, it follows that moisture-related expansion and contraction may not respond linearly to changes in RH. However, it cannot be assumed that dimensional change is directly correlated to the plastic's moisture content, as there can be other factors influencing how a plastic object may deform, such as the presence, amount, and type of plasticizer; the objects' initial dimensions; or its composite nature.

To successfully care for collections, it is important to understand the dimensional "cost" of raising or lowering the surrounding RH conditions. One way to measure this relationship is to monitor the dimensional change of a material as it moves from a desiccated state to a saturated state. This is undertaken in the laboratory by systematically raising the RH and measuring the change in length of the material as it is held under tension at a very low, constant stress. Figure 5.5(1a) illustrates this for different CTA film formulations, namely non-plasticized and plasticized, and with and without a gum-arabic-based paint layer, as found on animation artworks. This shows how the dimensions of the different films change as a percentage of their original dimension (strain) as the environment around the films change.

Another way to visualize this information is to calculate how much strain a material experiences in proportion to each step change in RH, which is determined by dividing the strain by the change in RH. This gives the humidity expansion coefficient (HEC) and shows "how much it costs in strain" for the sample to equilibrate in a new environment. Figure 5.5(1b) shows the HEC for the same CTA formulations. A completely flat HEC line indicates a perfectly linear behavior and illustrates that the amount of strain is constant per percentage RH change. Regions of the HEC graph that are curved indicate that percentage changes in RH do not generate a constant response in the hygroscopic plastic material. Investigating this problem in cycles can also inform how repeated changes in RH levels affect the material's response over time.

Figure 5.5(1b) shows that the presence of plasticizers reduces the RH-related expansion in both the painted and non-painted CTA films. The presence of very hygroscopic paint significantly increases the pace of change at the extremes of the RH scale, seen as a downward and upward curve on the graph. This is diminished across the 30–50% RH region. As noted elsewhere, changes in dimension are not necessarily a problem. However, knowing that various materials will have different responses to their environment helps to inform environmental management of collections. In the case of these painted CTA films, some materials respond differently to drying than to humidification, and in some cases, they have a predictable linear response during both humidification and drying across a region as broad as 10–50% RH. However, it is apparent that the drying process more often results in a

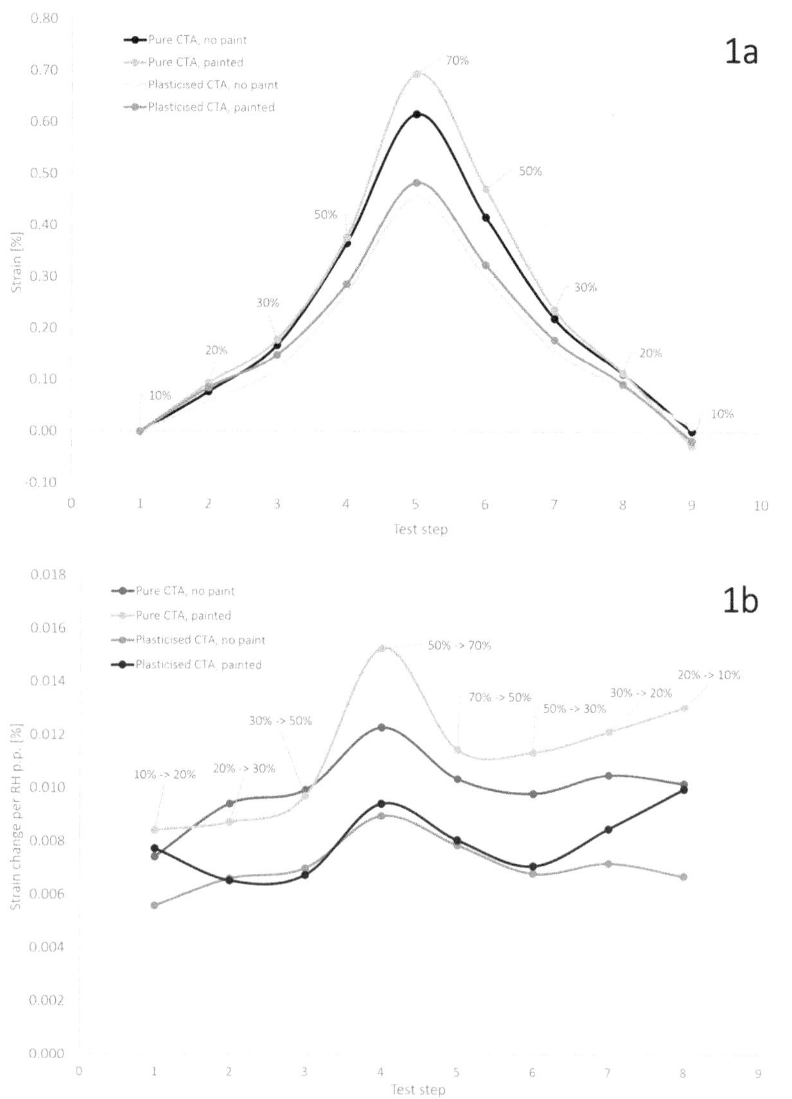

Figure 5.5 (1a) Moisture-induced expansion (strain) at 20°C for different cellulose triacetate film plasticized and non-plasticized, and with and without a gum-arabic-based paint layer at indicated levels of RH; (1b) calculated hygroscopic expansion coefficient representing the change in strain per percentage change in relative humidity at indicated transition steps.

greater rate of change than is seen during the humidification process, which may prove informative for dynamic environmental management.

Being conscious of those factors can greatly help smart environmental management of plastics collections and determine what level of fluctuation can be safely tolerated. Different polymers will exhibit characteristic HEC responses that are directly associated with their underlying composition. Therefore, more research is needed to establish these relationships for the range of plastics housed in collections.

Pollutants

Pollutants can come from the plastics themselves, as discussed in Chapter 4, as well as from the surrounding environment.

Pollutant Sources

Environmental pollutants can be from outdoor pollutants as well as indoor sources such as furniture, carpet, paint, case components like sealants and gaskets, and the collections. Typically, the "new smell," as in the smell of a new car, new carpet, or new paint, is caused by VOCs or semi-VOCs. Depending on concentrations and air exchanges, these pollutants can build up particularly within cases. Formic acid and acetic acid, which are emitted from wood and some plastics, among other sources, tend to be the most prevalent internal pollutants that damage collections (Tetreault 2003). So, if you have certain plastics in wood cases, there is a high potential for acids emitted from both. If the plastic is not off-gassing, the wood likely is, and this will be adsorbed by the plastic, leading to damage. And if the plastics in a wood case are off-gassing acids themselves, a highly acidic environment can form. Cases with high air exchange rates could lessen this concentration. However, a leaky case could permit additional pollutants to enter from the broader room. As a reminder, higher temperatures accelerate chemical reactions; a survey of storage in Danish museums found that indoor pollution concentrations markedly increased as temperatures rose in summer, so there can be seasonal pollutant risks (Rhyl-Svendsen 2019).

Pollutant Monitors

It is possible to monitor indoor air pollution, but most monitors are made for outdoor use or target levels tied to health concerns. Because pollutants that damage collections may be at lower concentrations or different from those that affect health or are commonly studied in the outdoor environment, specialized knowledge is needed to choose the right instruments for monitoring pollutants in collection spaces (Brimblecombe 2019). Passive samplers such as dosimeters and diffusion tubes have been available for collection care for decades, though there are factors that limit their effectiveness (Hatchfield 2019). Despite not being overly complicated to use, they tend to remain the purview of institutions with conservation scientists or conservators on staff but could be worth pursuing to determine the type of indoor pollutants.

Since it is not common for most institutions to monitor indoor pollution levels, it may be helpful to look at outdoor pollutant data, which is often readily available through online sources for many locations. This can be a helpful starting point to know what pollutants are in the general area that may be entering collection spaces (Brimblecombe 2019).

If it is suspected that the environment inside cases themselves contains pollutants, there are options for testing. If scientific instrumental analysis is available, headspace-solid-phase microextraction-gas chromatography/mass spectrometry (HS-SPME-GC/MS), as discussed in Chapters 3 and 4, is an option for studying emitted degradation products and has been used to determine cross-infection for a variety of plastics (Curran et al 2014).

More accessible options for determining the possible presence of some pollutants within cases include A-D (acid-detecting) Strips and the Oddy Test. If acids are suspected, A-D Strips are an easy-to-use, low-cost monitoring option that is covered in-depth in Chapter 7. The Oddy Test is an accelerated aging test that uses silver, copper, and lead coupons to gauge whether a storage or case material is appropriate for collection use. A further discussion of the Oddy Test is in Chapter 6. Using the concept of metal sensitivity to pollutants from the Oddy Test, the same metal coupons – silver, copper, and lead – can be set out in cases for longer-term, unaccelerated monitoring, as outlined in Chapter 7.

Pollutant Mitigation

If pollutants are coming from the external environment, determine if the building envelope can be better sealed. If the building is equipped with an HVAC system, check how much the current filters block and what the HVAC system can manage. If the air is recirculated as opposed to pulled from the outside, are those filters scavenging out common indoor pollutants, namely acetic and formic acid? The Minimum Efficiency Reporting Value (MERV) is an international filter rating system with the higher number, the more pollutants filtered. If an HVAC is not designed for it, a high MERV number filter can put too much stress on the system. Also, ensure that filters are routinely changed. Clogged filters reduce air flow, which will strain the system. The schedule for filter changes will depend on the recommended maintenance for the equipment in place as well as external influences based on location, with dustier locales and areas experiencing wildfires likely requiring more frequent filter changes.

If indoor pollutants are the issue, determine if sensitive plastics are better protected in cases. If the issues are from the cases themselves, consider the discussed monitoring options to determine the pollutant source, which may be from wood, paint, gaskets, or another component. If the offender is a collection item, should it be removed, or should other collections be removed from the case and the problem object left in isolation in the case? Could increasing ventilation of the case lower the pollutant level? Could passive adsorption options like activated carbon/charcoal or zeolites be used to scavenge acids or VOCs? If plastics are present, is there a risk of plasticizers leaching out if these are used, as described in Chapter 4? Could rehousing the collections in buffered archival papers and corrugated boards, like blueboard, be enough to counter released acids? If this is the route, periodic monitoring with A-D Strips or longer-term monitoring with metal coupons, as covered in Chapter 7, could be used to flag when the capacity of these materials to scavenge pollutants is exhausted.

For plastics vulnerable to oxidation, like hard rubber and PUR ether foam, low-oxygen or anoxic storage may be considered. There is collection care information readily available for setting up anoxic storage, so it will not be outlined here. However, there are a few points to keep in mind:

- Case-level anoxia typically involves an inert gas like nitrogen or argon to purge oxygen, which can be expensive and hard to maintain
- Individual anoxic storage in bags uses oxygen scavengers, like Ageless, which have exothermic reactions, meaning heat is released
- If the bag or case is not well-sealed, oxygen levels will rise, and scavengers will quickly deplete
- Anoxic containers have to be sealed, so collection access is limited

A study from The Costume Institute found that storage with Ageless may actually contribute to the deterioration of plasticized PVC, polyester, silicone, and 3D-printed nylon, possibly from the scavengers being too aggressive (Garcia, Hays, and Scaturro 2020). This study also found that if anoxic storage is desired but if it is not possible with an inert gas like argon, rather than using an oxygen scavenger, storage in an airtight container may be beneficial.

Sidebar 5.4 Anoxic Storage

Christopher Mazza

Materials like PUR and PVC, commonly used in the production of clothing and accessories, are unstable and under normal conditions may degrade over the course of several decades. One initial indication of degradation can be a loss of flexibility. Within a museum collection, exhibiting objects made of these materials becomes complicated when this loss of flexibility impacts mounting and display. In 2016, the Costume Institute acquired a dress (2016.13) by designer Iris van Herpen composed of PUR ether embedded with iron filings. To extend the displayable life of the dress, conservators and collections staff devised a two-part storage approach. Recognizing that PUR ether degrades in the presence of oxygen, the primary strategy aimed to minimize the dress's exposure to oxygen by placing it in an anoxic storage enclosure. As a secondary measure, to preserve the dress's shape within the enclosure, it was placed on a fully supportive storage mount. When the object is exhibited in the future, the storage mount will allow for one of the following options: Should the dress retain flexibility, it can be removed from the mount and dressed on a mannequin or form. However, should the dress lose flexibility, there remains the alternative of displaying the object directly on its storage mount.

Environments for Composite Objects

What do you do if an object contains more than one type of plastic and each of them has different environmental recommendations?

The Plastic Identification Tool (n.d.) offers the following options:

- compromise: find the middle ground between recommendations
- survival of the weakest: chose the recommendations for the most susceptible material
- majority: the most occurring material determines the conditions
- value-based: the character-determining material determines the conditions

Your institution does not need to establish a set approach to always use for composite objects because they each are unique. So, the decision that works well for one object composed of multiple plastics may not be the best route for another. Identifying the types of plastics, their needs and vulnerabilities, and the reason it was collected should inform each care decision.

Key Points for Influencers on Plastic Condition: Environment

Environmental factors of light in the visible, UV, and IR spectrums, temperature, RH, and pollutants all impact plastics, but in some instances, the impact can be more pronounced. Knowing what plastics are in collections is important for knowing environmental sensitivities. But the environment can only be changed if it is understood, which is why monitoring is important. Results from monitoring can then be used to target mitigation options, all with the goal of lessening negative influences of the environment on plastics.

References

Brimblecombe, Peter. 2019. "Air Quality, Monitoring, and Management." In *Preventive Conservation: Collection Storage*, edited by Lisa Elkin and Christopher A. Norris, 429–445. New York, NY: Society for the Preservation of Natural History Collections.

CAMEO, the Conservation and Art Materials Encyclopedia Online. 2022. "Materials Database." Museum of Fine Arts, Boston. https://cameo.mfa.org/wiki/Category:Materials_database

Cole-Parmer. n.d. "Traceable Humidity Card with Calibration." Cole-Parmer. Accessed April 23, 2024. https://www.coleparmer.com/i/traceable-humidity-card-with-calibration-6-pk/0331306

Curran, Katherine, Alenka Možir, Mark Underhill, Lorraine T. Gibson, Tom Fearn, and Matija Strlič. 2014. "Cross-Infection Effect of Polymers of Historic and Heritage Significance on the Degradation of a Cellulose Reference Test Material." *Polymer Degradation and Stability* 107: 294–306. https://doi.org/10.1016/j.polymdegradstab.2013.12.019

Garcia, Kaelyn, Marina Hays, and Sarah Scaturro. 2020. "Fashion's Plastic Problem: Preventive Conservation for Synthetic Materials." Paper presented at the American Institute for Conservation 48th Annual Meeting, Virtual, Summer 2020.

Gaylord Archival. 2024. "Maximum Humidity Indicator Cards." Gaylord Archival. www.gaylord.com/Environmental-Control/Temperature-%26-Humidity-Monitors/Maximum-Humidity-Indicator-Cards-%285-Pack%29/p/MXC56789

Hatchfield, Pamela. 2019. "Evaluating Materials Used for Collection Storage." In *Preventive Conservation: Collection Storage*, edited by Lisa Elkin and Christopher A. Norris, 677–703. New York, NY: Society for the Preservation of Natural History Collections.

Himmelstein, Paul, Scott Rosenfeld, and Steve Weintraub. 2019. "Illumination for Collection Storage." In *Preventive Conservation: Collection Storage*, edited by Lisa Elkin and Christopher A. Norris, 215–228. New York, NY: Society for the Preservation of Natural History Collections.

IPI (Image Permanence Institute). 2024. "FilmCare.org, Storage Suitability." Image Permanence Institute, Rochester Institute of Technology. www.filmcare.org/storage_categories

Ledoux, Nicole, Georgina Rayner, Susan Costello, and Angela Chang. 2023. "Preventative Conservation, Treatment and Technical Study of Plasticized Poly(vinyl Chloride) Multiples by Joseph Beuys." *Studies in Conservation*, 68 (3): 343–356. https://doi.org/10.1080/00393630.2022.2033520

McCormick-Goodhart, Mark H. 2003. "On the Cold Storage of Photographic Materials in a Conventional Freezer Using the Critical Moisture Indicator (CMI) Packaging Method." www.wilhelm-research.com/subzero/CMI_Paper_2003_07_31.pdf

MoDiP. 2022. "Caring for Synthetic Garments." Arts University Bournemouth. www.modip.ac.uk/projects/curators-guide-synthetic-garments/curatorial-process/caring-synthetic-garments

Plastic Identification Tool. n.d. "INFO: Preventive Conservation." Cultural Heritage Agency: Ministry of Education, Culture, and Science. Accessed April 23, 2024. https://plasticidentificationtool.nl/info/Preventive-conservation

Rhyl-Svendsen, Morten. 2019. "Organic Acids in Confined Storage Areas." In *Preventive Conservation: Collection Storage*, edited by Lisa Elkin and Christopher A. Norris, 432–433. New York, NY: Society for the Preservation of Natural History Collections.

Shashoua, Yvonne. 2008. *Conservation of Plastics: Materials Science, Degradation and Preservation.* Oxford: Elsevier/Butterworth-Heinemann.

Shashoua, Yvonne. 2014. "A Safe Place, Storage Strategies for Plastics." *Conservation Perspectives, the GCI Newsletter* 29 (1) Spring 2014. www.getty.edu/conservation/publications_resources/newsletters/pdf/v29n1.pdf

Simione, Frank P. 2019. "Low Temperature Storage." In *Preventive Conservation: Collection Storage*, edited by Lisa Elkin and Christopher A. Norris, 473–489. New York, NY: Society for the Preservation of Natural History Collections.

Talas. 2024. "Humidity Cards." *Talas*. www.talasonline.com/Humidity-Cards

Tetreault, Jean. 2003. *Airborne Pollutants in Museums, Galleries and Archives: Risk Assessment, Control Strategies and Preservation Management*. Ottawa: Canadian Conservation Institute.

van Oosten, Thea. 2022. *Properties of Plastics: A Guide for Conservators*. Los Angeles: Getty Conservation Institute.

Weintraub, Steve. "Specialized Macroclimates and Microclimates: Options for the Control of Temperature, Relative Humidity, and Pollutants." In *Preventive Conservation: Collection Storage*, edited by Lisa Elkin and Christopher A. Norris, 449–471. New York, NY: Society for the Preservation of Natural History Collections.

Williams, R. Scott. "Plastic Storage Products." In *Preventive Conservation: Collection Storage*, edited by Lisa Elkin and Christopher A. Norris, 753–779. New York, NY: Society for the Preservation of Natural History Collections.

6 Influencers on Plastics

Material Choices

The materials and forms of support that are chosen for storage, exhibition, and transport can affect how art and artifacts age and that include plastics. Since many of these support materials are plastics themselves, it is important to understand their properties and material interactions. This chapter will cover plastics commonly used in collection care applications and highlight how even though many are in the same plastic family, their forms, uses, vulnerabilities, and effects on collections can differ.

Collections are typically housed in a lot of white plastics such as Ethafoam, Volara, and Tyvek. This may be because white is the default color available for these products but that itself may be because suppliers know that collection care professionals want to be able to see dust, detached fragments, and evidence of pests which are all more obvious against a white background and since they are not colored with dyes, white products avoid the risk of staining materials they are in contact with (Arenstein, Goldberg, and Milroy 2019). However, it is interesting to consider that carbon black can be added to plastics as ultraviolet (UV) stabilizers to retard photochemical reactions from penetrating beyond the surface, therefore, potentially increasing longevity (Williams 2019). So, while black plastics are not commonly used for storage supports, in some instances they may be worth considering if made from a class of plastic usually deemed safe for collections (Table 6.1) and pass the Oddy Test for permanent use (see below).

Table 6.1 is an alphabetical listing of plastics usually considered safe for collection use. The contents of Table 6.1 are discussed further in this chapter for a fuller understanding of the materials, uses, and limitations. Changes in formulations and other factors can render a product unsuitable so it is important to purchase from a reputable supplier that understands museum needs, ask questions about products, be engaged with collection-focused discussion forums in case colleagues report issues, reference results of material testing, such as the Oddy Test, before purchasing (discussed below), and periodically check collections for adverse changes that may be caused by housing materials.

It is important to consider that the plastics discussed in this chapter are from non-renewable resources and have a large carbon footprint. With growing sustainability efforts, the use of these plastics will likely decrease. The more sustainable

DOI: 10.4324/9781003489658-6

Table 6.1 Plastics Usually Considered Safe for Collections Use

Plastic Type	*Abbreviation*	*Uses*	*Names*
Polyamide	PA	Fiber, sheet, film	Nylon, Dartek
Polycarbonate	PC	Sheet, panel, glazing	Lexan
Polyethylene (high density)	HDPE	Padding, lining, support, interleaving	Ethafoam, Tyvek, Volara, Plastazote, Hollytex (*also as polyester*)
Polyethylene (low density)	LDPE	Containers, sleeves, enclosures, sheeting	Tupperware, Sterilite (*also as PP*)
Polyethylene-Polypropylene copolymer		Enclosures, frames, backing, support	Coroplast, Polyflute, Correx
Polyethylene terephthalate	PET or PETE	Sheet, sleeve; batting/fill	Mylar, Melinex, polyester
Polymethyl methacrylate	PMMA	Sheet, panel, glazing	Acrylic, Plexiglas, Perspex, Lucite, Acrylite
Polypropylene	PP	Container, board, sleeves	Tupperware, Sterilite (*also as LDPE*)
Polystyrene	PS	Container, cushioning	Styrofoam
Polytetrafluoroethylene	PTFE	Sheet, support, gasket	Teflon, Gore-Tex
Silicone	SI	Coating, sealant, gasket	Silastic

materials that hopefully replace them, however, will need to be assessed for potential interactions with collections; the content in the "Resources to Aid with Material Choices" section in this chapter should be able to inform these decisions as well.

Polyamide

Polyamide (PA) is a family of plastic that is likely best known by the tradename Nylon. It can be made into fibers, bristles, films, sheets, as well as three-dimensional forms. PA may find its way into collection care applications as: fishing line to secure vulnerable elements, the hooks in Velcro, cable ties (also known as zip ties or Ty-Rap), Nylon rivets to secure Coroplast boxes or tray corners, and as Dartek film for wrapping paintings or other works with vulnerable surfaces (Arenstein, Goldberg, and Milroy 2019; Preparation, Art Handling, Collection Care Information Network (PACCIN) 2025; Plastic Identification Tool n.d.). PA is also being used more in 3D printing applications, as covered in Sidebar 2.2.

PA is hygroscopic, so will take on moisture in high RH environments and release that moisture when the RH drops thus expanding and contracting with RH fluctuations, as discussed in Sidebars 5.2 and 5.3 (Plastic Identification Tool n.d.). Studies have found that the pigment titanium dioxide accelerates photo-oxidation, especially for PA fibers, contributing to their yellowing (MoDiP 2022; Plastic Identification Tool n.d.). So, it is best to keep PA polymers in stable RH and protect them as much as possible from visible and UV light.

Polycarbonate

Polycarbonate (PC) is a transparent plastic that comes in films and sheets and may be used for either flat or curved glazing so can have similar applications as polymethyl methacrylate (PMMA). PC is best known by the tradename Lexan. PC has high impact resistance but is vulnerable to scratching (Conservation and Art Materials Encyclopedia Online (CAMEO) 2022). PC can block UV light but doing so turns it yellow. To counter this, stabilizers and antioxidants are added during manufacture but since these will eventually be used up, PC will yellow over time (Plastic Identification Tool n.d.). Because of this vulnerability, exposure to sources of UV, such as sunlight, should be avoided. Be aware that PC becomes brittle at low temperatures (Plastic Identification Tool n.d.). A potential health concern with PC is that bisphenol A (BPA), an endocrine disruptor that mimics estrogen, may have been used in manufacture and can be released as PC degrades (Sidebar 2.4) (Curran et al 2014). The use of BPA in plastics intended for infants and toddlers, such as bottles and formula packaging, are banned in many countries but its use may be permitted in other products.

Polyethylene

Polyethylene (PE) is the most common plastic today (Plastic Identification Tool n.d.). It is also typically the most frequently used plastic in collection storage. PE comes in two forms: high density and low density. As evident in Table 6.1, PE has become ubiquitous in 21st-century collection care as bags, containers, sheets, and cushioning supports. It is almost impossible to enter collection storage and not find PE products, often used in tandem. For example, picture a cavity mount. The base material for the recessed support is likely Ethafoam, a stable, closed-cell foam that is waterproof and chemically resistant. Though stable and supportive, cut Ethafoam edges are abrasive so they are usually covered with either Volara, a smooth closed-celled flexible PE foam, or Tyvek, which are PE sheets made of spunbonded continuous filaments that are permeable but resistant to wetting. Even though Ethafoam, Volara, and Tyvek have different physical properties and are very different products, they are all **high-density polyethylene (HDPE)**. A marker for all HDPE is higher molecular weight with a tightly packed, linear chemical structure making the final product strong and opaque (van Oosten 2022).

Listed below are forms and uses for HDPE in collections with the tradename or common name in parentheses, any notable concerns, and typical uses:

HDPE Foams (Ethafoam, Volara, Backer rod, Plastazote): As PE foams age, they can become brittle, may not bounce back to shape, and may produce an acrid odor though damage to materials is not usually associated with PE foams (Williams 2019).

- *Ethafoam*: planks of varying thicknesses. Shock absorption in transport, cut into support wedges or for cavity packing.
- *Ethafoam, Volara*: sheets of varying thickness. Soften contact points for objects, boxes, and shelf liners. Thin sheets can depress under weight.
- *Backer rod or foam rod*: rods of varying diameters, support for rounded items like vases, tray dividers. Note that polyurethane (PUR) backer rod also exists but should NOT be used.
- *Plastazote*: gaskets for cases.

HDPE Fibers (Hollytex):

- Nonwoven, lightweight, porous, strong, smooth, inert fabric used for support, filtering, and interleaving. Can also be made using polyester fibers.

HDPE Sheets (Tyvek):

- Dust covers (permeable so will not block gaseous pollutants)
- Textile supports (Sidebar 8.1)
- Envelopes and sleeves (Sidebar 2.1)
- Liners for crates (permeable so will not block gaseous pollutants)
- Sewn into pillow shapes and stuffed (often with polyester batting)
- Interleaving
- Wrapping

Since Tyvek is commonly used in storage and transport, it is helpful to know a bit more about it. On its website, DuPont describes Tyvek as "Feels like paper, but it's not paper. Feels like fabric, but it's not fabric" (2024). It comes in two forms: hard and soft. Hard Tyvek, designated as Type 10, is smooth, stiff, and more paper-like (Sidebar 2.1). In collection care, the hard Tyvek style 1025D is most often used for interleaving, wraps, and support slings (Arenstein, Goldberg, and Milroy 2019). Soft Tyvek, indicated as Type 14, is more like fabric since it is soft, washable, able to be sewn, and has a more obvious weave-like structure on one side and is glossy on the other. For collections, style 1443R is used to sew supports, dust covers, and as linings (Arenstein, Goldberg, and Milroy 2019). The glossy side is made with an anti-static coating intended to repel dust while the textured side tends to hold onto dust. For some collections, it is standard to wash Tyvek in distilled water or just in cold water without detergent prior to use in order to remove or reduce surface coatings (Global Conservation Forum 2020).

Though Tyvek has historically been considered stable for extended use with collections, it is important to note that concerning interactions have been reported in scientific studies and observations from use. One study on the effect of plastics on cellulose noted that Tyvek emitted acetic acid at room temperature, even though deionized water was wiped across the surface to remove the anti-static coating, indicating that the culprit was likely PE as opposed to the surface treatment (Curran et al 2014). Since PE is vulnerable to thermal oxidation at room temperature (20°C or 68°F), that may have played a role in this finding and could put materials in the vicinity at risk (Williams 2019). Several conservators reported Tyvek issues on a variety of materials ranging from copper alloys to paintings to plastics from "ghosting" of the embossed weave pattern of soft Tyvek imprinting on surfaces from contact either during transport or in long-term storage. Rather than caused by a chemical interaction, the imprints seem to usually be the result of pressure, abrasion, or the formation of a microclimate that softens surface coatings of the encased objects making them more vulnerable to impressions (Global Conservation Forum, 2020). So, direct prolonged contact of Tyvek with some collections comes with risks in some instances. Remember also that since Tyvek is permeable, volatile organic compounds (VOCs) either emitted from a plastic or from other materials, such as wood, will pass through it.

Low-density polyethylene (LDPE) has a lower molecular weight than HDPE and has loosely packed, branched molecules that permit flexibility and allow light to pass through (van Oosten 2022). For collections, LDPE is mostly found as containers, bubble-wrap film, sheeting, and zipped-top storage bags and is usually evident by its waxy feel and translucency.

Listed below are forms and uses for LDPE in collections with the tradename or common name in parentheses, any notable concerns, and typical uses:

LDPE Containers (Tupperware, Sterilite)

- Various sizes, often with lids. Many companies that make LDPE containers also make them in polypropylene (PP) so check for a Resin Identification Code (Chapter 3) to determine if it is LDPE or PP.

LDPE Film (Bubble Wrap, MarvelSeal)

- Bubble Wrap: packaging with bubbles to cushion during transport. For best impact protection, face bubbles in but know that the bubble pattern can imprint so avoid direct contact with objects by using an interleaving layer. Some brands are coated with polyvinyl chloride (PVC) or polyvinylidene chloride (PVDC) which may form hydrochloric acid (CAMEO 2022). Therefore, do NOT use Bubble Wrap for long-term storage.
- MarvelSeal: vapor barrier used to protect from moisture and/or off-gassing, such as from wooden crates or shelves, and for sealed packages and custom enclosures. MarvelSeal 360, made of aluminum foil sandwiched between nylon film and layers of LDPE film, is the most frequently used version for collection care. MarvelSeal 2175 (formerly MarvelSeal 470) consists of aluminum foil, PE, and PP and is also sometimes used with collections; it is more easily sealed than MarvelSeal 360 but less resistant to puncture (Arenstein, Goldberg, and Milroy 2019; CAMEO 2022).

LDPE Sheeting

- Rolls of varying thicknesses. Used to protect against potential or active leaks and to encase materials sensitive to water damage or pests, such as large textiles in open storage. If used for wrapping, since it is translucent rather than transparent, it can be difficult to fully see what is being stored. Static can build up, causing loss of small fibers or loosely bound materials.

LDPE Sleeves and Bags

- Various sizes and thicknesses. Bags typically have zipped-tops used to individually store small objects. May yellow over time. Edge seams may fail putting the object at risk of dissociation especially if catalog information is labeled on the bag/sleeve or on a tag within the bag rather than on the object itself. Do NOT use food-grade LDPE bags with collections.

Yellowing, as seen in Figure 6.1, can obscure contents within LDPE enclosures but may also be a sign of weakening of the polymer, likely from oxidation degradation that can be driven by exposure to visible and UV light as well as heat, even at room temperature (Williams 2019). Antioxidants delay this process but will become exhausted and, somewhat ironically, oxidation of antioxidants can be responsible for yellowing (Stuart et al 2023). Oxidation can produce colored by-products called chromophores. During this process not only can yellowing result but molecular mass decreases, thus reducing polymer strength. In bags this weakening may cause edge failure, putting contents at risk for dissociation (Sidebar 6.1). Butylated hydroxytoluene (BHT) can be used as an antioxidant, particularly in food-grade LDPE bags, and as BHT volatilizes, it can be adsorbed by other materials nearby resulting in discoloration (Williams 2019). As a result, do not use food-grade LDPE bags for collection storage.

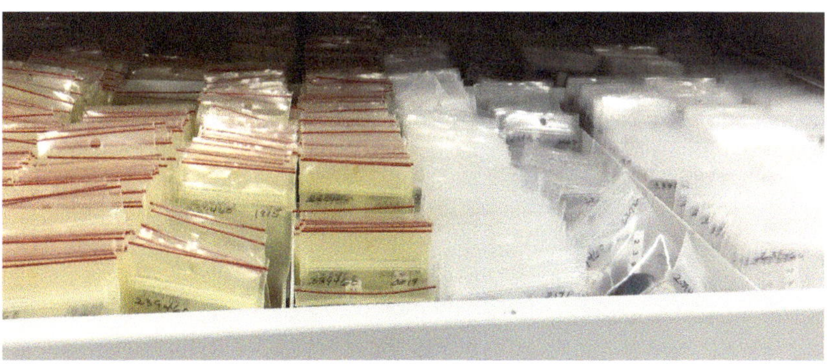

Figure 6.1 Older yellowed LDPE bags alongside newer LDPE bags to store small objects.

Despite these considerations, LDPE is still commonly used in collection care and the practice is still considered acceptable for most collections. While yellowing is not always an issue, it is important to remember that by-products from discoloration can sometimes be a risk to nearby materials and there is concern that objects in direct contact may interact with degradation products which has the potential to affect the object itself over time and/or interfere with scientific analysis thus reducing research potential, as explored in Sidebar 6.1 (Stuart et al 2023).

In the article "Safe Storage? An Assessment of Polyethylene for the Storage of Heritage Objects," the authors draw attention to "cleanroom" PE bags as a potential alternative since they are manufactured to meet more stringent requirements (Stuart et al 2023). The adoption of these purer PE enclosures in collection care will undoubtedly be dependent on cost, availability, results of future research, and, ideally, actual versus perceived need. Or in other words, if "cleanroom" PE plastics prove to be safe for storage but are expensive rather than deciding that rehousing into them is out of reach, collection stewards could target which collections to store in them by weighing the needs of specific objects, curatorial priority, or a combination of factors.

It is vital to remember that there are some materials, such as PVC, that should NEVER be stored in PE. One study of the accelerated aging of 30-year-old PVC sheets found that when in contact with LDPE, plasticizers migrated from the PVC into the LDPE. The rate of this migration was quick with the LDPE gaining 2% mass in the first month while the PVC lost 2% of its weight in that period (Royaux et al 2020). Therefore, PVC should not be stored in contact with PE in any form, whether a LDPE sheet or bag or a HDPE container. Rather, storage in non-adsorbent containers like glass have proven to be better for PVC preservation (Shashoua 2003).

Polyethylene and Polypropylene Copolymer

Since PE and PP polymers have similar chemical structures consisting of carbon and hydrogen atoms (polyolefin chemical family) and have comparable appearances and uses, it is not surprising that they can be used together as a copolymer. In collection storage, a copolymer of PP and PE may be found as corrugated plastic sheets made into trays and used for backing and support panels, often under the tradenames Coroplast, Polyflute, and Correx (Pasiuk 2004; CAMEO 2022). These can come in different colors, which may be familiar in daily life as small yard signs used to promote a political candidate or to proclaim that a neighbor graduated. When used outdoors, additives such as colorants and UV blockers are likely present. But in collection storage, the translucent option is usually most common but that does not mean that additives are absent. Coatings, anti-static agents, and flame retardants may be present and could affect collections. Additionally, be careful if housing materials with friable surfaces because static charges can accumulate causing loosely bound materials and fibers to shed. Some collection caretakers have reported that the corrugated channels can harbor insects or be prone to mold growth after exposure to water, such as after a leak.

Polyethylene Terephthalate

PE terephthalate (PET or PETE) is likely most familiar to those in collection care as a transparent film often described in product catalogs as archival polyester or by the tradenames Mylar or Melinex. It is typically used for lining or encapsulation and comes in rolls of different thicknesses that can be cut to size or purchased in pre-made sleeves and sheets. Versions of rolled film with a silicone (SI) coating are sometimes used when there are concerns of sticking, such as with a weeping plastic (Sidebar 10.1). However, there may be risks in using PET in contact with plasticized PVC. When in contact with PET, a research team found that during accelerated aging experiments PVC discolored, turning dark brown in areas, and the PET eventually stuck to the PVC (Royaux et al 2020). Though PET did not cause plasticizer migration from PVC, as the researchers found with PE, the PVC plasticizer degraded at a faster rate when in contact with PET, which is a concern. It is worth noting that accelerated aging tests do not always predict real-world aging and most in the field still consider Mylar/Melinex to be an inert option for collection care. And as discussed in Sidebar 5.1, storage of plasticized PVC wrapped in PET at low temperatures may have potential benefits.

PET is reported to be able to block some UV but damage potential still exists from light (Pasiuk, 2004). Static build-up can be very pronounced, putting friable collections at risk for loss especially when being rolled, unrolled, or taken in and out of a sleeve or wrapping. PET can be a good blocker for acidic vapor which may be helpful in some collections with off-gassing plastics. For example, PET could be used to line the backs and sides where open shelves meet to prevent emitted acids from degrading plastic from migrating to adjacent collections but still allow for ventilation. Remember encasing an off-gassing plastic in PET would trap acids, which can accelerate degradation of plastics like cellulose nitrate (CN) and cellulose acetate (CA), so is not recommended.

PET can also be made into polyester fibers and used as batting stuffed into a stable outer fabric for rehousing applications where soft support is needed, such as for padding out textiles or shoes.

Polymethyl Methacrylate

PMMA is often referred to as acrylic or by the tradenames Plexiglas, Perspex, or Lucite. Historically, it is interesting to know that these three were all registered around 1933 with Plexiglas being patented in Germany, Perspex in England, and Lucite in the United States (Plastic Identification Tool n.d.). PMMA is generally used for glazing in frames and cases and has the advantage of being half the weight of glass. It can be treated to prevent glare, block UV, and cut down static. It has also been extensively used for object mounts and book cradles, though many are moving away from PMMA for new mounts in favor of more sustainably minded options like metal. PMMA is considered an inert, stable plastic but some grades are vulnerable to crazing from environmental stressors and can be damaged by some cleaning products so take heed of cleaning guidelines from the manufacturer

(Williams 2019). Because PMMA is often used for cases in public spaces, scratches can form which can be difficult to reduce.

Polypropylene

PP is the second most commonly produced commercial plastic today, behind PE (CAMEO 2022). PP is chemically similar to PE and they can be hard to distinguish but, generally, PP is slightly less dense so can be more transparent than PE of the same thickness and is glossier and more rigid than PE (CAMEO 2022; Plastic Identification Tool n.d.). In collections, PP may be found as vapor barriers, protective sleeves for papers and photos, and various-sized containers. Since both PP and PE can be made into translucent containers by the same manufacturers, check for a Resin Identification Code to distinguish between the two (Chapter 3). PP can also be made into fibers, waxes, and lubricants, and be used for cushioning or insulation.

As with PE, PP is vulnerable to thermal oxidation at room temperature and to photodegradation by exposure to visible and UV light which can cause yellowing and embrittlement (Williams 2019). While yellowing may not be harmful, it can obscure contents and be visually disturbing to the extent that some may think the collection is not being well cared for. If enclosures become brittle, they may fail which could lead to damage or loss.

In a study to determine if plastics cross-infected cellulose, three PP samples had no effect but one PP sample "...displayed a significant impact on the degradation of cellulose" (Curran et al 2014). The study determined that the aggressive PP sample contained a flame retardant that likely accelerated the degradation of that PP polymer, serving as a reminder that though plastics may be from the same family, additives and other alterations to formulation can affect how they age. In this case, it meant that a plastic that is usually deemed stable and safe for use with collections actually was not.

Polystyrene

In its rigid form, polystyrene (PS) sounds tinny when tapped and is used for petri dishes and as packaging, such as for CD cases (van Oosten 2022). PS can also be a hard, lightweight foam best known by the tradename Styrofoam. PS foam is used for impact cushioning and insulation and is usually white and sounds squeaky when rubbed. PS has environmental sensitivities to visible and UV light, oxidation, moisture, temperature, and some solvents like acetone which will blanch rigid clear PS and burn through PS foam (see solubility parameter discussion in Chapter 8) (Plastic Identification Tool n.d.). Degradation causes yellowing, embrittlement, crumbling, crazing, and cracking (van Oostern 2022). There have been concerns of styrene off-gassing from the polymer that could affect nearby varnishes and coatings; however, this is not considered a risk with modern PS manufacturing (Williams 2019). PS can be sensitive to plasticizers that weep out from other plastics so while considered to be a stable plastic, its use for in-contact storage with other plastics comes with risk (Plastic Identification Tool n.d.).

Polytetrafluoroethylene

Polytetrafluoroethylene (PTFE) is generally categorized as a high-performance plastic meaning that it has special properties that make it costlier and not generally found in commercial goods (van Oosten 2022). PTFE is likely best known by the tradename Teflon. For collection care, Teflon may be found in gaskets and as Teflon tape or plumber's tape which is used in varying widths to secure elements together, such as identification tags, or as a method for smoothing over abrasive contact points like the cut edges of Ethafoam (Arenstein, Goldberg, and Milroy 2019).

Gore-Tex is a waterproof, breathable fabric that may be familiar from winter outerwear. It is made of a PTFE film laminated to a nonwoven fabric known as Hollytex that is made from either polyester or PE fibers (CAMEO 2022). While not generally utilized in storage, Gore-Tex is used in conservation treatments for localized humidification to relax wrinkles and curling, to aid in stain removal, and to soften adhesives.

Silicone

SI, sometimes referred to as silicone rubber because of its rubber-like elastomer qualities, is flexible with a smooth, oily feel in its three-dimensional form and may be most familiar as cookware. SI oil may migrate out of three-dimensional products causing the surface to become tacky which makes dust likely to stick. Exuded SI oil can stain porous surfaces they are in contact with so interleaving, such as with Mylar or Melinex, is recommended. As SI oil weeps out, the flexible properties of SI will lessen (Sidebar 10.1). Despite these issues SI is considered a stable plastic if it is a "manufactured product" meaning that it is factory-made under controlled conditions as opposed to mixed by individuals which introduces the risks of incorrect proportions or inadequate curing which may cause issues as the SI ages (Williams 2019).

SI may be utilized as caulks and gaskets for museum cases, both for storage and exhibition (Hatchfield 2019). Though SI is generally considered stable, because formulations can vary, even among the same manufacturer, the particular product should be evaluated before use, such as with the Oddy Test (see below). For example, platinum-cured SI and alkoxy-based one-part SI sealants may be inert but other SI caulks and gaskets may emit ammonia or acids, especially as they cure (Hatchfield 2019). If these off-gassed products are trapped within a case, the collection materials within may negatively react. Even if deemed stable, it is important to allow curing time after application before collections are placed within a silicone gasketed case.

SI coated Mylar or Melinex may be employed in collection care when there are concerns of sticking (Sidebar 10.1). Another option in some applications may be SI release paper which is paper impregnated or coated with a small amount of SI to make the paper water resistant, non-adsorbent, and prevent it from sticking to most surfaces, as discussed in Sidebars 4.1 and 8.1.

Sidebar 6.1 Storage Plastics and Sample Viability in Archeological Research

Julie Unruh and Wendi Field Murray

Storage environment plays a central role in determining sample viability for materials analyses – and from an analytical perspective, plastics can be tricky. In the context of one type of analysis, (ancient DNA, for example), PP plastic might be the preferred material for sample storage, while for others (i.e., gas chromatography/mass spectrometry (GC/MS)), plastic contamination of samples might compromise the reliability of results. Some analyses, like X-ray fluorescence (XRF), may be indifferent to plastics altogether. To further complicate the issue, the same sample might be ideal for multiple types of analyses, and not all researchers agree on appropriate storage for the samples they analyze. As instrumental analyses in collections research continue to increase (and as collection managers continue to make storage decisions for collections), it behooves us to better understand the effects of plastic storage materials on analytical results, and how storage decision-making in the repository can, in some instances, be a make-or-break proposition for the viability of collections samples.

Broadly speaking, plastics may affect analytical viability three ways. They may alter the object chemistry via transfer of foreign substances: for example, plastics may leach additives or manufacturing residues (Stuart et al. 2023) that are transferable to other objects via surface contact or handling. They may produce environmental microclimates, both desirable and detrimental: for example, storage with limited air exchange can helpfully exclude oxygen or moisture, or can harmfully promote fungal growth. And finally, they can have physical effects – hard plastic containers can encourage breakage; plastics may accumulate damaging static electricity; and plastic storage containers may degrade, promoting dissociation and loss.

Below are short lists of analytical techniques that may be impacted by storage plastics. The lists are not comprehensive. Notable omissions include various types of chromatography, scanning electron microscopy/energy-dispersive spectrometry (SEM/EDS or EDX), and X-ray diffraction (XRD).

Analyses with outstanding concerns as to whether lab-quality plastic storage is permitted or is susceptible to contamination by plastics:

- **Fourier-transform infrared spectroscopy (FTIR)**
  - Julie Arslanoglu, Research Scientist, The Department of Scientific Research, The Metropolitan Museum of Art, New York. Personal communication. February 21, 2024.
  - Molly McGath, Research Scientist, The Mariners' Museum and Park. Personal communication. February 6, 2024 and April 16, 2024.

  - Adriana Rizzo, Research Scientist, The Department of Scientific Research, The Metropolitan Museum of Art, New York. Personal communication. February 28, 2024.
- **Gas chromatography/mass spectrometry (GC/MS)**
  - Julie Arslanoglu, Research Scientist, The Department of Scientific Research, The Metropolitan Museum of Art, New York. Personal communication. February 21, 2024.
  - Wendy Matthews, Associate Professor in Archaeology, Department of Archaeology, School of Archaeology, Geography and Environmental Science, University of Reading. Personal communication. August 3, 2023.
  - Reber (2018)
  - Adriana Rizzo. Research Scientist, The Department of Scientific Research, The Metropolitan Museum of Art, New York. Personal communication. February 28, 2024.
- **Organic residue analysis (an umbrella term for combined analytical techniques)**
  - Joseph G. Barabe, Senior Research Microscopist, Barabe & Associates LLC. Personal communication. July 20–21, 2023.
  - Kathleen J. Birney, Associate Professor of Archaeology, Wesleyan University. Personal communication. July 21, 2023.
  - Patrick E. McGovern, Scientific Director, Biomolecular Archaeology Project and Adjunct Professor, Anthropology, Penn Museum. Personal communication. July 26, 2023.
  - McGovern (2019)
  - Oudemans and Erhardt (1996)
  - Christina Warinner, John L. Loeb Associate Professor of the Social Sciences and the Sally Starling Seaver Associate Professor at the Radcliffe Institute, Harvard University. Personal communication. February 1, 2024 and March 22, 2024.
- **Radiocarbon dating**
  - Megan Anne Conger, Assistant Research Scientist, Center for Applied Isotope Studies, University of Georgia. Personal communication. February 13, 2024.
  - Carla Hadden, Director, and Alex Cherkinsky, Senior Research Scientist, University of Georgia Center for Applied Isotope Studies, and PI of the Radiocarbon Dating Lab at the UGA Center for Applied Isotope Studies, respectively. Personal communication. July 31, 2023.
  - Yates, Smith, and Bertuch (2015)
- **Stable isotope analysis**
  - Megan Anne Conger, Assistant Research Scientist, Center for Applied Isotope Studies, University of Georgia. Personal communication. February 13, 2024.

- Fraser, Meier-Augenstein, and Kalin (2008)
- Vaiglova (2023)

Analyses affected by non-breathable microclimates:

- Every analysis researched was reported to be adversely affected by fungal growth with the exception of XRF.
- Trapped humidity may encourage movement of material components such as plasticizers or soluble salts through each other with resulting contamination (Molly McGath, Research Scientist, The Mariners' Museum and Park. Personal communication. February 6, 2024 and April 16, 2024).

Analyses affected by physical effects:

- Any technique may be affected by breakage, dissociation, and loss of sample material due to storage container degradation.
- Static electricity can physically modify lightweight materials or impede safe removal of samples from storage containers (Joseph G. Barabe, Senior Research Microscopist, Barabe & Associates LLC. Personal communication. March 22, 2024; Adriana Rizzo, Research Scientist, The Department of Scientific Research, The Metropolitan Museum of Art, New York. Personal communication. March 22, 2024.

Changes wrought by plastics in the storage environment can be invisible but permanent. For many analyses, pre-treatments and other contingencies can help scientists navigate a known source of contamination or degradation; however, researchers must be informed about historical exposures in order to deploy those corrective measures. For that reason, collection care professionals should maintain records of the storage histories of objects and make them available to analysts. Additional controlled studies are needed to understand the effects of plastic on sample viability, and to build collection care standards that maximize collections' research potential.

Fire Risks of Plastic Material Choices

If a museum professional was asked to name a fire risk for plastics, most would likely bring up CN film, as discussed in Chapter 4. But all petroleum-based plastics come with a high fire load though the temperature at which they burn, the degree to which they fuel a fire, and the toxins they release as they burn will vary by plastic type. Considering how much plastic is used in museums, particularly in storage rooms, fire concerns are real. A controlled fire in a simulated storage room carried out by the Smithsonian Institution and the National Institute of Standards and Technology found that compared to other materials plastics release more heat, produce more smoke, melt, and spread further with the following results for specific types of storage plastics (Stewart 2019):

- PE sheeting melted to objects on open shelving
- Tyvek disintegrated early in the fire
- Nomex and Kevlar-wrapped objects fared better than those in Tyvek

Though there is a clear risk of plastics in a fire, it would be difficult and perhaps not even advisable to remove them from use in storage. For example, if an institution is prone to leaks and collections are on open shelving, covering the shelving with PE sheeting is a sound decision since the risk of damage from leaks is so high. Like most aspects of collection care weighing the probability of risks, in this case the higher likelihood of leaks given past occurrences versus a fire starting, could mean that using PE sheeting is not only a reasonable decision but also the responsible one until there is a more permanent fix.

Considerations for Accessioned Plastics

For most rigid three-dimensional plastics, standard storage materials like blueboard, a buffered archival corrugated board discussed in Chapter 4, is a solid choice. However, with some plastics in collections there is a need for special considerations for materials used to house them. Plastics that are weeping or are plasticized and have the potential to weep, may benefit from using SI release Mylar or Melinex or SI coated paper to prevent the plastic from sticking at points of contact or transferring degradation products, as discussed in Sidebars 4.1, 8.1, and 10.1. As a reminder, accelerated aging studies found that PET aggressively darkened plasticized PVC and hastened the degradation of the plasticizer (Royaux et al 2020). Therefore, using SI paper may be a better option in some instances.

Some flexible plastics were made to inflate. These tend to be novelty items like beach balls and pool floats or could have some functionality like a blowup chair. Because of the need for inflatables to be flexible and the desire to have them be affordable, many are made of PVC. When deciding how to store these items, keep in mind that the plastic would have a high degree of plasticizers to allow for flexibility to inflate and deflate. Over time, as the plasticizers are lost, the plastic will get tacky and become brittle. For collection care, this means that inflatables should be stored blown up in order to hold their shape. If they are stored deflated, in time that is how they will have to remain, making their interpretation limited. The question of how to keep the shape of inflatables periodically comes up in the field, as evidenced by searching "inflatables" in the Global Conservation Forum, formerly known as the ConsDistList. In recent years when this question has been raised, concerns of inflating with air were noted because trapped moisture can result in mold growth on the interior (Walthew 2018). The majority of conservators and conservation scientists who responded on the forum to the post "Looking for advice on storing inflatable PVC object" recommend using nitrogen to blow up inflatable PVC objects because nitrogen will slow the degradation mechanism dehydrochlorination, as described in Chapter 4 (Global Conservation Forum 2023). If filled with either air or nitrogen, eventually the gases will escape either through the seams, leaky valves, or permeation through the plastic itself. So, it is necessary to monitor and periodically re-inflate these collection items. Because inflatables are plasticized, SI release sheets should be used at points of surface contact to prevent sticking. Care should be taken when handling these collections to both protect the objects and the handler. As plasticizers migrate to the surface they can be absorbed through the skin so proper handling precautions are needed (Sidebar 2.4). And remember, it can be difficult to know by sight when a once flexible plastic turns rigid and is vulnerable to handling-induced stress.

Resources to Aid with Material Choices

Resources on specific products, either from the manufacturer or gathered for museum-based applications, may provide information that can be utilized for collection care decisions.

An example of a product resource is **Safety Data Sheets (SDS)**, formerly known as Material Safety Data Sheets (MSDS), which can be used to get insight into potential risks that a product may pose to collections, especially when combined with product literature from the manufacturer (Hatchfield 2019). In the SDS, the sections that provide information about the physical and chemical properties of the material that are the most helpful in assessing material risks are (Hatchfield 2019):

Section 1: *Identification.* Identifies the chemical and recommended uses.
Section 2: *Hazard(s) Identification.* Identifies chemical hazards and warnings.
Section 3: *Composition/Information on Ingredients.* Identifies the ingredient(s) in the product including impurities and stabilizing additives.
Section 9: *Physical and Chemical Properties.* Identifies physical and chemical properties of the product.
Section 10. *Stability and Reactivity.* Information on reactivity hazards and chemical stability.

Since SDS are geared for human safety, not the safe preservation of collections, you will have to extrapolate information for collection care applications. For instance, if the SDS for a product points out that it is acidic or is an oxidizing agent, it may be a collection risk. Also know that proprietary information or trade secrets can be withheld from the SDS so there can be unknown variables in the product.

There are online resources designed to help with choosing materials for use with collections, whether for storage, exhibition, or transport. Since 1997, when it began at the Museum of Fine Arts, Boston, the electronic database **Conservation and Art Materials Encyclopedia Online (CAMEO)**, provides technical information on materials, terms, and techniques connected to the creation and preservation of art (CAMEO 2023). The Materials Database portion of the site "…contains chemical, physical, visual, and analytical information on over 10,000 historic and contemporary materials used in the production and conservation of artistic, architectural, archaeological, and anthropological materials" (CAMEO 2022).

From the CAMEO Materials Database landing page, entries can be searched from an alphabetical list or by typing a specific name in the search field. If you need a product to fit a category of use but are not sure what is available, the database can also be searched using broader terms like "interleaving" or "foam." The contents in the results pages can slightly differ but generally the following sections, or similarly named ones, show in the entry along with relevant images and any analytical spectra that may be helpful:

1 Description
2 Synonyms and Related Terms
3 Applications
4 Personal Risks
5 Collection Risks
6 Environmental Risks
7 Physical and Chemical Properties
8 Working Properties
9 Forms/Sizes
10 Resources and Citations

CAMEO is an accessible first stop when researching a material or wondering about its applications and is kept up-to-date with a "last edited" time stamp at the bottom of each entry.

Of these CAMEO entries, there are currently 185 tied to the Materials Working Group (MWG) which was created under the Collection Care Network, since rebranded as the Preventive Care Network (PCN), of the American Institute for Conservation (AIC). The entries connected to the MWG are materials used for exhibition, storage, and transport both now and in the past. Though older materials may no longer be recommended, they may still be found in collections and thus need to be understood. A strength of the entries connected to the MWG initiative is that they also pull from information posted on the website for the Preparation, Art Handling, Collections Care Information Network (PACCIN) and include tips for working with the products (CAMEO MWG 2023).

The MWG developed a wiki called **"Choosing Materials for Storage, Exhibition & Transport"** that is hosted by AIC and can help collection caretakers work through material choices (Arenstein et al 2022). The site is formatted with the following sections which may be helpful in determining what materials to use with plastics as well as what plastic materials to use:

1 Why is the Choice of Materials Important?
2 What are the Factors to Consider when Choosing Materials?
   - 2.1 Physical Characteristics and Working Properties
   - 2.2 Financial, Logistical, & Operational Accessibility
   - 2.3 Health and Safety

- 2.4 Environmental Sustainability
- 2.5 Chemical Stability

3 How do I Prioritize these Factors?
- 3.1 Assessing Vulnerability
- 3.2 Assessing Threat from Products
- 3.3 Assessing Context

4 How Do I Choose What to Use?
- 4.1 Select from lists of already tested materials
- 4.2 Research and test materials on your own
- 4.3 Work directly with a collection care professional

5 Resources

Another resource is the **"Materials Testing Results"** wiki on which contributors compile the results of primarily Oddy Testing carried out on materials used in case, container, and crate fabrication (AIC Wiki 2024). As mentioned in Chapter 5, the Oddy Test is an accelerated aging test used to determine if materials are suitable for permanent or temporary use with collections or are unsuitable. The Oddy Test entails the following:

- Enclosing a sample of the material in question in a container with three polished and cleaned metal coupons (silver, copper, and lead) and a vial of water to create a high humidity environment (about 95–100% RH), since humid conditions will drive some reactions.
- For 28 days the sealed container is placed in an oven at 60°C (140°F) since heat accelerates chemical reactions.
- After the test, any changes to the metals are noted. The metal sensitivities are as follows:
  - Silver (Ag) for sulfur compounds
  - Copper (Cu) for chlorides, oxides, and sulfur compounds
  - Lead (Pb) for organic acids, aldehydes, and acidic gases
- If any of the metals react, the material fails for long-term, permanent use.
- If a slight reaction on a metal is observed, the material may be deemed acceptable for temporary use (no more than six months).
- Testers may interpret results differently, making the Oddy Test subjective.

Keep in mind that formulations of the materials tested may change over time so a material from one batch or supplier may pass for permanent use while a sample from another shipment or time period may fail. Also, because the test depends on interpretation of changes, it is subjective. What one person denotes as unsuitable another may deem okay for temporary use. This disparity plays out with many of the products in Table 6.1 when the "Materials Testing Results" wiki is searched. Despite these limitations, the Oddy Test is regarded as a helpful gauge for material choices and the Oddy Test wiki serves a role in assisting collection choices when targeted testing is not feasible either because of staff, budget, or time constraints (AIC Wiki 2024).

Key Points for Influencers on Plastics: Material Choices

Plastics accessioned into collections can have unique issues to consider, such as those illustrated by inflatable plastics. The choices made for storage, which for many objects will fully encompass the majority of time in a museum, as well as for exhibition, which in some instances can be several decades, are vitally important for preservation. Since many frequently used storage and display options are composed of plastics, it is important to make informed choices. Those making decisions have to understand what materials choices are available, their limitations, and risks they may bring with them, such as increased fire load and potential interactions with materials. Online resources such as CAMEO and those developed by the MWG initiative and hosted on the AIC Wiki can help inform these decisions.

References

AIC Wiki. 2024. "Materials Testing Results." American Institute for Conservation. www.conservation-wiki.com/wiki/Materials_Testing_Results

Arenstein, Rachel Perkins, Lisa Goldberg, and Eugenie Milroy. 2019. "Support and Rehousing for Collection Storage." In *Preventive Conservation: Collection Storage*, edited by Lisa Elkin and Christopher A. Norris, 641–675. New York, NY: Society for the Preservation of Natural History Collections.

Arenstein, Rachael Perkins, Mary Coughlin, Lisa Elkin, Lisa Goldberg, Jennifer K. Herrmann, Rebecca Kaczkowski, Julia Sybalsky, Margalit Schindler, Elena Torok, and Joelle Wickens. 2022. "Choosing Materials for Storage, Exhibition & Transport." American Institute for Conservation. www.conservation-wiki.com/wiki/Choosing_Materials_for_Storage,_Exhibition_%26_Transport

CAMEO (Conservation and Art Materials Encyclopedia Online). 2022. "Materials Database." Museum of Fine Arts, Boston. https://cameo.mfa.org/wiki/Category:Materials_database

CAMEO (Conservation and Art Materials Encyclopedia Online). 2023. "Conservation & Art Materials Encyclopedia Online." Museum of Fine Arts, Boston. https://cameo.mfa.org/wiki/Main_Page

CAMEO MWG (Conservation and Art Materials Encyclopedia Online, Materials Working Group). 2023. "Fact Sheets on Exhibit & Storage Materials." https://cameo.mfa.org/wiki/Category:MWG

Curran, Katherine, Alenka Možir, Mark Underhill, Lorraine T. Gibson, Tom Fearn, and Matija Strlič. 2014. "Cross-Infection Effect of Polymers of Historic and Heritage Significance on the Degradation of a Cellulose Reference Test Material." *Polymer Degradation and Stability* 107: 294–306. https://doi.org/10.1016/j.polymdegradstab.2013.12.019

DuPont. 2024. "What Is Tyvek." DuPont. www.dupont.com/tyvekdesign/design-with-tyvek/why-tyvek.html

Fraser, Isla, Wolfram Meier-Augenstein, and Robert M. Kalin. 2008. "Stable Isotope Analysis of Human Hair and Nail Samples: The Effects of Storage on Samples." *Journal of Forensic Sciences* 53 (1): 95–99. https://doi.org/10.1111/j.1556-4029.2007.00607.x

Global Conservation Forum. 2020. "Tyvek "Ghosting"." December 17, 2020.

Global Conservation Forum. 2023. "Looking for Advice on Storing Inflatable PVC object." November 30, 2023.

Hatchfield, Pamela. 2019. "Evaluating Materials Used for Collection Storage." In *Preventive Conservation: Collection Storage*, edited by Lisa Elkin and Christopher A. Norris, 677–703. New York, NY: Society for the Preservation of Natural History Collections.

Ledoux, Nicole, Georgina Rayner, Susan Costello, and Angela Chang. 2023. "Preventative Conservation, Treatment and Technical Study of Plasticized Poly(vinyl Chloride) Multiples by Joseph Beuys." *Studies in Conservation*, 68 (3): 343–356. https://doi.org/10.1080/00393630.2022.2033520

McGovern, Patrick E. 2019. "Guide to Organic Residue Analysis." Philadelphia: Biomolecular Archaeology Project. University of Pennsylvania Museum of Archaeology and Anthropology, Philadelphia, PA.

MoDiP (Museum of Design in Plastics). 2022. "Nylon." Arts University Bournemouth. www.modip.ac.uk/projects/curators-guide-synthetic-garments/featured-fibres-and-materials/nylon

Oudemans, Tania F. M., and David Erhardt. 1996. "Organic Residue Analysis in Ceramic Studies: Implications for Conservation Treatment and Collections Management." *Studies in Conservation* 41 (sup1): 137–142. https://doi.org/10.1179/sic.1996.41.Supplement-1.137

PACCIN (Preparation, Art Handling, Collection Care Information Network). January 15, 2025. "Dartek." www.paccin.org/content.php?72-Dartek

Pasiuk, Janet. 2004. "Safe Plastics and Fabrics for Exhibit and Storage." *Conserve O Gram* 18/2. National Park Service. www.nps.gov/museum/publications/conserveogram/18-02.pdf

Plastic Identification Tool. n.d. "Plastics." Cultural Heritage Agency: Ministry of Education, Culture, and Science. Accessed April 19, 2024. https://plasticidentificationtool.nl/

Reber, Eleanora A. 2018. "Gas Chromatography-Mass Spectrometry (GC-MS): Applications in Archaeology." In *Encyclopedia of Global Archaeology*, edited by Claire Smith, 1–17. Cham: Springer International Publishing: 2958. https://doi.org/10.1007/978-3-319-51726-1_340-2

Royaux, Adeline, Emilande Apchain, Isabelle Fabre-Francke, Nathalie Balcar, Gilles Barabant, Clémentine Bollard, Bertrand Lavédrine, Odile Fichet, and Sophie Cantin. 2020. "Conservation of Plasticized PVC Artifacts in Museums: Influence of Wrapping Materials." *Journal of Cultural Heritage* 46: 131–139. https://doi.org/10.1016/j.culher.2020.07.002

Shashoua, Yvonne. 2003. "Effect of Indoor Climate on the Rate and Degradation Mechanism of Plasticized Poly(vinyl Chloride)." *Polymer Degradation and Stability* 81 (1): 29–36. https://doi.org/10.1016/S0141-3910(03)00059-4

Stewart, Josh. 2019. "Flammability of Rehousing Materials." *AIC News* 44 (2): 25–27. March 2019.

Stuart, Barbara, Sharon Wong, Rosemary Goodall, Alice Beale, Cancy Chu, Petronella Nel, Holly Amin-Jones, and Paul Thomas. 2023. "Safe Storage? An Assessment of Polyethylene for the Storage of Heritage Objects." *Studies in Conservation* 68 (6): 669–678. https://doi.org/10.1080/00393630.2022.2092702

Vaiglova, Petra. 2023. "Sampling Archaeological Remains for Isotopic Analyses." In *Field Sampling for Laboratory Analysis in Archaeology*, edited by Evi Margaritis, Artemios Oikinimou, Efthymia Nikita, and Theo Rehren, 139–146. www.academia.edu/103088284/Field_Sampling_for_Laboratory_Analysis_in_Archaeology

van Oosten, Thea. 2022. *Properties of Plastics: A Guide for Conservators.* Los Angeles: Getty Conservation Institute.

Walthew, Jessica. 2018. "Inflatable Furniture." *Global Conservation Forum*, November 30, 2018.

Williams, R. Scott. "Plastic Storage Products." In *Preventive Conservation: Collection Storage*, edited by Lisa Elkin and Christopher A. Norris, 753–779. New York, NY: Society for the Preservation of Natural History Collections.

Yates, Andrea Bettina, Andrew M. Smith, and Fiona Bertuch. 2015. "Residue Radiocarbon AMS Dating Review and Preliminary Sampling Protocol Suggestions." *Journal of Archaeological Science* 61 (September): 223–234. https://doi.org/10.1016/j.jas.2015.06.011

7 Low-Cost, On-Site Monitoring of Plastics

This chapter focuses on options for easy-to-use, low-cost passive monitoring that can be carried out by staff without specialized scientific training. The options outlined do not require scientific instrumentation, collection of samples to send to a lab, or an outside consultant. Each method has its pros and cons, which are covered to empower collections staff to make informed decisions about which options to utilize, what the results likely indicate, and how that can inform decisions for plastics care.

The monitoring options covered in this chapter are:

- Metal coupons
- A-D Strips
- Cresol Purple or Cresol Red indicators
- Ozone Test sticks

Metal Coupons

The metal coupons typically utilized for long-term monitoring mimic those from the Oddy Test: silver, copper, and lead (Chapter 6). These metals are deployed because of their sensitivities to gaseous pollutants that can be present in indoor environments and affect collection materials:

- Silver (Ag) for sulfur compounds
- Copper (Cu) for chlorides, oxides, and sulfur compounds
- Lead (Pb) for organic acids, aldehydes, and acidic gases

For better reactivity and more reliable results, use as pure form of the metals as possible (close to 100%). Thin metal sheets, sometimes called foil, can be bought from scientific and metal suppliers or even jewelry-making suppliers and cut to size with clean scissors. The surfaces of the coupons need to be fresh, without a developed oxide layer or patina, and without contamination from oils and salts from hands. Wearing accelerant-free, powder-free nitrile gloves during coupon

DOI: 10.4324/9781003489658-7

preparation and whenever handling the coupons is important. Some in conservation have noted that nitrile gloves, specifically those not denoted as accelerant-free, can play a role in corroding metals they are in contact with, likely from leaving behind chlorides and sulfur residues from processing. But accelerant-free gloves can have issues as well. As with any mass-produced plastic, the formulation of nitrile gloves will vary by manufacturer and even the same manufacturer will change ingredients based on cost and availability of ingredients. Thus, any recommendation of one brand of glove over another is just a snapshot of results at the time of testing. That said, it is helpful to review the tests carried out at the Metropolitan Museum of Art to better understand why some nitrile gloves may have handling concerns with metals (Stephens et al 2022). To avoid any potential issues from nitrile gloves, it is best to handle the coupons after cleaning using clean metal tweezers.

Protocol for metal coupon preparation can vary among institutions, as outlined on the American Institute for Conservation (AIC) "Oddy Test Protocols" Wiki (2024), but the following method should be accessible to most collection care staff:

- Wear accelerant-free nitrile gloves
- Set out acid-free paper to work on
- Clean metal tweezers and scissors with acetone
- Polish silver and copper coupons to remove surface oxidation or soiling using either a glass bristle brush or fine grit micromesh sandpaper. Use a different brush or sandpaper for each metal. Polish under ventilation, which can be achieved by setting up a high efficiency particulate air (HEPA) vacuum. Wear proper personal protective equipment (PPE): dust mask, safety goggles, and lab coat.
- Do NOT polish lead coupons because of health hazards
- Cut metal coupons to size
- Clean metal coupons with acetone either by submerging them or using cotton swabs under ventilation and with PPE.
- Put metal coupons in place, usually near the object in a closed storage or display case, using cleaned metal tweezers. Do not put metals and plastics in contact when monitoring.

Once the metals coupons are in place, if possible, check weekly for changes. If no changes appear after several weeks can be altered to a monthly schedule (Coughlin 2011). The metal coupons are helpful for longer-term monitoring since they can be left in place. If a metal reacts, the likely culprit can be flagged for further monitoring in situ or in isolation and then possibly selected for deaccessioning (Chapter 9). For the purposes of this book, the assumption will be that the cause of metal corrosion is an accessioned plastic or plastic(s) used in case construction or in the storage or display housing but remember that other components like

Table 7.1 Metal Coupon Sensitivity to Problem Plastic Degradation Products

Plastic	*Possible Emission*	*Metal Coupon*
Cellulose Acetate	Acetic acid	Lead
Cellulose Nitrate	Nitrous oxides/nitric acid	Copper
Hard Rubber	Sulfur	Silver
Polyurethane Ester Foam	Acids	Lead
PVC	Hydrogen chloride/hydrochloric acid	Copper

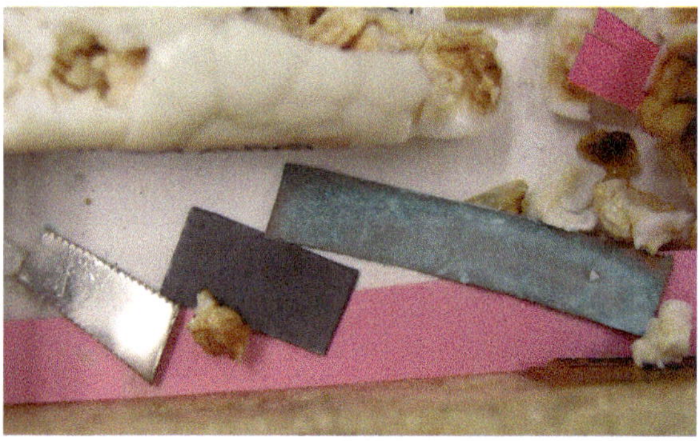

Figure 7.1 Monitoring severely degraded CN samples with silver, lead, and copper coupons over four weeks. As a strong oxidizing agent, the nitric acid corroded the copper after just one week. The bright pink in the image is reacted Cresol Purple indicators seen in Figure 7.3. NMAH 1983.0538.87/.90/.91. Photographed by author, Division of Medicine and Science, National Museum of American History, Smithsonian Institution.

adhesives, wood, and paint can off-gas and corrode metal coupons. Generally, a positive reaction for each metal appears as:

- Silver: dark gray, black tarnish
- Lead: gray or white corrosion either even or mottled. May also be a reddish-brown color
- Copper: green but may also be brownish or whitish

As with traditional Oddy Testing (Chapter 6), using metal coupons for long-term monitoring is subjective. Because they can accelerate the degradation of each other, ensure that the plastic and metals are not touching. If the types of plastics are known, possible products that may be emitted as they degrade can allow targeted deployment of metal coupons, as seen in Table 7.1. However, if the type of plastic is uncertain or there are concerns about other pollutant sources, it is best to use all three metals for monitoring, as shown in Figure 7.1.

Since metals can corrode when exposed to degrading plastics that emit reactive compounds, it is a good idea to look at the hinges, handles, screws, and nails used in cabinet construction as well as other nearby accessioned metal objects that may be kept in the same enclosure. If it is difficult to explain the cause of the metal corrosion and a plastic could be a candidate for off-gassing, it would be helpful to isolate the plastic in question and monitor it with an appropriate monitoring method from this chapter to determine if it is the culprit.

Acid-Detecting (A-D) Strips

Acid-detecting (A-D) Strips became available from the Image Permanence Institute (IPI) in 1995 to monitor acetic acid (vinegar syndrome) from degrading cellulose acetate (CA) film but were quickly repurposed by the field to monitor a variety of other collection materials such as wood and other plastics (Nicholson and O'Loughlin 1996). The strips start as a deep cobalt blue and turn through shades of green to yellow in the presence of acidic vapor in the surrounding air with the quicker and more yellow the reaction indicating a higher acid concentration, as seen in Figures 7.2, 7.4, and 7.5 (Image Permanence Institute (IPI) 2022).

Figure 7.2 Celanese Celluloid Corp. CN color sample from 1944 showing uneven aging, including level of acidic off-gassing after three days of monitoring with A-D Strips. NMAH 1983.0538.93. Photographed by author, Division of Medicine and Science, National Museum of American History, Smithsonian Institution.

IPI has a five-step color scale from 0 to 3 that equates color change to the degree of acidity in CA film, with the higher the number (yellower) the more acid present:

0 = blue
1 = bluish green
1.5 = green
2 = greenish yellow
3 = yellow

To determine color change, the test strip is compared to a reference card mailed with the strips. In the past, IPI included a pencil with the color scale printed around it. However, the green color was not lightfast making the use of the pencil for color comparisons potentially unreliable if the pencil was not stored in the dark (Hackney 2016). So, always use the reference card mailed with new strips as opposed to the pencil you may have had for years. If the reference card is lost, an unused strip from the package can be used to compare how far the exposed strip shifted from the original dark blue.

The indicator dye used on the paper to make A-D Strips is bromocresol green buffered with sodium hydroxide and essentially measures a pH range of 5.4–3.8 with the lower numbers equating to increased acidity (Hackney 2016). Since the pH scale is 0–14 with 7 being neutral, readings of a lower number value are more acidic. The dye becomes less and less blue in the presence of increased acid which is why A-D Strips go from blue through shades of bluish green to green to greenish yellow to yellow.

Because the strips are sensitive to a range of acids, the plastic in question needs to be isolated in a sealed container to allow for any acidic off-gassing to concentrate, thus, aiding in the detection by the strips. Since the build-up of degradation products can accelerate autocatalytic degradation in some plastics, as discussed in Chapter 4, the duration of sealed A-D testing should be limited to a few weeks at most. Additionally, the reliability of the A-D Strips after about a month of exposure has been brought into question in collection-focused studies (Hackney 2016; Townsend, Hackney, and Kearney 2023).

By testing in an enclosure, the source of acidic pollution, whether that means determining which object among several or which component of a single object is the issue, and how far the acid is spreading, can be better monitored. The strip should be placed near the object but not in contact with it in order to prevent any staining if the dye gets wet (IPI 2022). If this proves difficult, one study noted the possibility that "...miniature neodymium magnets, stacked several high and with the strips held between two of them, could assist in maintaining the position of the strips..." but warn that if the object is large, it would be wise to deploy multiple strips because the area that looks the most degraded may not actually be emitting the most acid (Townsend, Hackney, and Kearney 2023).

The time it takes for A-D Strips to react depends on certain variables. IPI notes that very heavy acidic off-gassing will change the strips in a few minutes at standard museum conditions. If less acid is being emitted, it will take at least 24 hours to register but most people deploy the strips for longer which is especially important if plastics other than CA film are being monitored since different acidic degradation products from other plastics can take longer to build-up and register on the strips. If monitoring occurs at lower temperatures or lower humidity, it will take longer for the strips to react (IPI 2022). The assumption is that lower temperatures or lower relative humidity (RH) are slowing down the rate of chemical degradation. So, if an object is in low-temperature storage, A-D Strips can still be used but they will need to remain in place for a longer time. High RH, between 60 and 90%, causes more pronounced color change while slight variations to RH in the range typically aimed for in a standard museum environment, in the study that was 45–60% RH, did not affect results (Hackney 2016). Temperatures that could be considered at the extreme ends of room temperature, 16.5°C (61.7°F) and 27.5°C (81.5°F), which may occur in poorly controlled interiors over the course of the year, should not impact the reaction of the A-D Strips (Hackney 2016).

A-D Strips are light-sensitive and will fade under ambient light after a few days (IPI 2022). So, testing is best done in the dark. If that is not possible, testing should not be planned for more than a few days, with three days likely being a reasonable amount of time to achieve reliable results (Townsend, Hackney, and Kearney 2023). Even in the dark, results from longer-term testing over several months may not be accurate because of color shifting but at standard room conditions for temperature and RH testing over several weeks is acceptable (IPI 2022).

When the period of monitoring ends, immediately evaluate the strips for a color change by assessing it next to the reference card sent with the strips or an unexposed new strip. Do the comparison even if it looks like no reaction occurred since sometimes the change is only obvious when the strip is next to a reference. Ambient light can be warm or cool and affects how the human eye interprets color. Those who work with collections may be familiar with this from colors standing out more or looking different under studio lighting for photography or under exhibition lighting compared to storage. Because of the effect of ambient lighting on color interpretation, IPI notes that the comparison of the strips to the reference card was designed to occur under cool-white fluorescent light (IPI 2022). When using the A-D reference card, note the corresponding number for the color change. IPI recommends the use of half measurements so that if a color does not quite line up, assume the worst and go up to the next closest level, making this the possible scale: 0, 0.5, 1, 1.5, 2, 2.5, and 3 (IPI 2022). It is important to remember that the reference card interpretation is only for CA film but can still be helpful for other monitoring. CA films that are in good condition may actually cause a blue-green reaction (between 0 and 1 on the reference card). For a result of 1.5 or higher, the CA film is at the point of autocatalytic degradation and should be flagged for duplication and/or frozen storage (IPI 2022).

The acids adsorbed by the strips will evaporate even after just a few minutes when removed from the acidic environment, changing to a bluish-green color. The strip will not revert back to its original blue because some of the reactive products on the strip are used up by exposure to acids (Hackney 2016). IPI advises that in order to retain reacted strips for comparison later on, they can be taped to a clear stable plastic sheet, such as Mylar or Melinex, and stored in the dark for several months (IPI 2022). While this may be helpful for reference, ensure that the clear tape itself is stable. The carrier and/or adhesive on some tapes can be acidic, causing the A-D Strip secured under it to react making moot the effort to preserve evidence of the strip's reaction to the plastic in question. So, test any tape used with A-D Strips to confirm they are inert. Or to avoid the potential for tape contamination, collected A-D Strips could be placed in Mylar or Melinex envelopes with the air flattened out, wrapped in aluminum foil, or encased in a combination of Mylar or Melinex and aluminum (Hackney 2016).

General Testing Protocol for A-D Strips for Small Individual Plastics

- Select an inert, transparent container such as a glass beaker or glass food storage container.
  - Lids for glass food storage are often made from plastic so if sealing the container with its lid, first monitor the lid with A-D Strips (put the A-D Strip in the container sealed with its lid). If no reaction after one week in the dark, safe to use for testing. For repeated use, know that the plastic lid may adsorb products from degrading plastics during testing and contaminate future tests.
- If using a glass beaker or container without its lid, seal the top with aluminum foil and wrap Parafilm around the edges to ensure a tight seal.
- Wearing nitrile gloves, place plastic in the chosen container.
- Using clean metal tweezers, place A-D Strip in the container.
- Avoid having the A-D Strip and plastic touch in case of dye transfer (the dye is soluble so can run if it gets wet and may stain).
- Seal the container and place it in the dark.
- Monitor for one week, ideally checking daily. If not possible, check after three to seven days.
- Can monitor for several weeks, but not longer than a month.
- Do not open the container if checking more than once or any off-gassed products will dissipate, affecting the test results.
- After the monitoring period, remove the A-D Strip from container and evaluate immediately by comparing it to the reference card or new, unused A-D Strip.
- If conducting subsequent tests, make sure to clean out and air out containers between use to prevent cross-contamination and false positive results.
- Do not reuse strips, even if they did not react.

General Testing Protocol for A-D Strips In Situ, Such as in a Case

- Using clean metal tweezers, place A-D Strip in case.
- Avoid having the A-D Strip and plastic touch in case of dye transfer (the dye is soluble so can run if it gets wet and may stain).
- Monitor after three to seven days (the assumption being that a case or container would have to be open to get close enough for daily checking which would lower the acid concentration and affect results).
- Can monitor for several weeks, but not longer than a month.
- After the monitoring period, immediately evaluate the A-D Strip by comparing to the reference card or new, unused A-D Strip.
- Do not reuse strips, even if they did not react.

According to IPI, A-D Strips that are sealed in their unopened original packaging remain viable for five years (2022). After opening, keep unused strips in the original packaging (PE zipped-top bag within a MarvelSeal pouch which can be folded over and secured shut) to prevent color shifting from light and reaction with ambient environmental pollutants. If the packaging is damaged or lost, unused strips can be stored by encasing in Mylar or Melinex envelopes and stored in the dark or wrapped in aluminum foil or a combination of the two as an added barrier from light and ambient pollutants (Hackney 2016). Store the strips in a cool environment, not exceeding 25°C (77°F), that is not too dry (IPI 2022).

Remember, plastic can be degrading but not be off-gassing acids. It could also be that any acidic products have already been emitted or that other non-acidic degradation products are being released that A-D Strips are not sensitive to. Because of this and since A-D Strips are only a snapshot of the moment of testing, deploying metal coupons for longer-term monitoring as described above can be helpful. Since plastic may be currently fine but may degrade in the future, plastics that are vulnerable to acidic off-gassing, especially problem plastics from Chapter 4, should be periodically monitored with A-D Strips. Early detection of emitted acids, which can occur even when the plastic looks fine, can help make storage and display decisions that are better for the plastic as well as collections around it.

Sidebar 7.1 Monitoring a Cellulose Acetate Necklace

Christopher Mazza

At the Costume Institute, A-D Strips are used to monitor a variety of fashion accessories made of CA (such as hair combs, sunglasses, and fans) for the chemical degradation often referred to as "vinegar syndrome" for the distinct vinegar smell of released acetic acid. These strips are used both as a survey tool for evaluating groups of objects in shared cabinetry and as a quantitative tool for determining the condition of individual objects. In the case of a recently acquired necklace (2018.752) by the designer Elsa Schiaparelli, composed of polychrome brass insects on a base of clear CA, the short-term strategy is to annually monitor for the early stages of degradation by sealing it in a polyethylene bag with an A-D Strip for 24 hours at an approximate temperature of 70°F (21.1°C). If the A-D Strip indicates that the CA is beginning to degrade, the necklace will be isolated to better meet its storage needs. This step will also protect other objects stored alongside the necklace, as degrading CA releases volatile organic compounds (VOCs) that can potentially damage nearby items. Ultimately, the long-term strategy for the necklace will be to move it into cool or cold storage as this can significantly extend the life of CA. This objective, however, is still in the planning phase as it requires rethinking current storage practices to prioritize material type over considerations such as object type and access.

Cresol Purple or Cresol Red Indicator Strips

Cresol purple or cresol red dye can be used to make indicators that change color in the presence of very strong acidic vapors emitted from cellulose nitrate (CN), as covered in Chapter 4 (Fenn 1995). As the emitted nitrous oxides change to nitric acid, the acidic vapor will react with the cresol red or purple indicator dye causing a characteristic color change:

- Cresol purple changes from yellow to purple or bright pink
- Cresol red changes from yellow to reddish purple

A positive reaction of either dye will take anywhere from one day to five days (Odegaard, Carroll, and Zimmt 2015). On severely degraded CN, however, a color change can occur in a few hours and in extreme cases within minutes, as seen in Figure 7.3. Generally, a quick and intense color change indicates that more acidic vapor is present and, therefore, the CN is further along on its degradation pathway.

The limitation of cresol red and purple indicators is that they are not available commercially so need to be prepared. *Material Characterization Tests for Objects of Art* has easy-to-follow instructions for prepping a 0.005% cresol red solution by adding 0.005 grams of cresol red to a mixture of 90-milliliter ethanol and 10-milliliter methanol (Odegaard, Carroll, and Zimmt 2015).

For cresol purple, there is a water-based version in which 0.04 grams of cresol purple is dissolved in 100 milliliters of deionized water in order to obtain a 0.04% w/v solution. Though the water-based version may seem friendlier to use, proper PPE including goggles, nitrile gloves, and protective clothing is still needed.

While the need to prep the solution in-house may be a disadvantage for some, it also presents the advantage of choosing the substrate material to apply the solution to and cutting it to size. For example, filter paper, yarn, string, and cheesecloth all

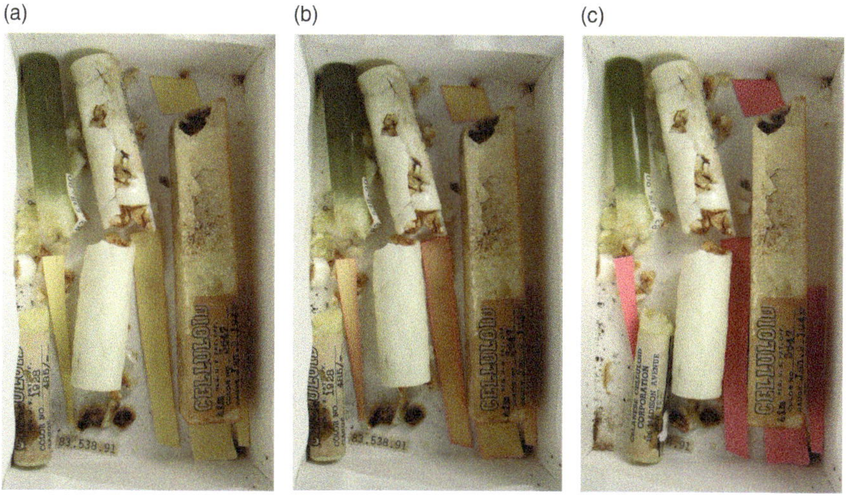

Figure 7.3 (a) Cresol purple with CN color samples from the early 1940s, (b) Reaction after a few minutes, (c) Reaction after a few hours. The metal grommets of the storage box are heavily corroded and the paper box itself is an acidic pH 3. NMAH 1983.0538.87/.90/.91. Photographed by author, Division of Medicine and Science, National Museum of American History, Smithsonian Institution.

can be used to take up the reactive dye and then be deployed for testing to see not only if acidic vapor is present but also how far it is spreading (Fenn 1995)

Since the acidic vapor is heavier than air, testing under the material in question as opposed to over it is recommended (Odegaard, Carroll, and Zimmt 2015). Though in heavily off-gassing situations, positive results can occur on indicators placed above a sample. Remember that the vapor can spread out beyond the CN object making it possible to contaminate nearby objects and be adsorbed by storage materials including other types of plastics.

Weaker acids like acetic and formic acid emitted from other types of degrading plastics do NOT cause a reaction with these indicator strips so they are only recommended for determining if nitric acid is being emitted from CN. Because cresol purple and red have a preferential sensitivity to acidic off-gassing from CN, in some instances, they can be used to identify it. For example, in Figures 7.4 and 7.5, the appearance and date of the eyeglasses mean that both CN and CA are possible candidates for the degrading plastic frames. Monitoring both with A-D Strips confirmed that acidic off-gassing is occurring; the addition of cresol purple strips helped distinguish between the two types of plastics. In Figure 7.4, since the A-D Strip reacted but the cresol purple stayed yellow (no reaction), the glasses were identified as CA, not CN. In Figure 7.5, both monitoring strips had positive reactions for acidic vapor, helping confirm those glasses as CN, not CA. Eventually, both pairs of eyeglasses completely fell apart and were flagged for deaccession (see Chapter 9).

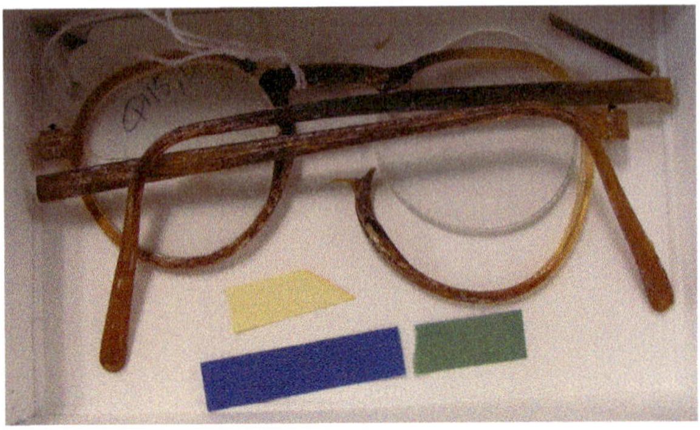

Figure 7.4 Eyeglasses monitored for three days. A-D Strip turned light green (positive reaction, compared to unused blue control strip). Cresol purple stayed yellow (no reaction). Monitoring helped confirm glasses as CA, not CN. NMAH MG.314637.12189. Photographed by author, Division of Medicine and Science, National Museum of American History, Smithsonian Institution.

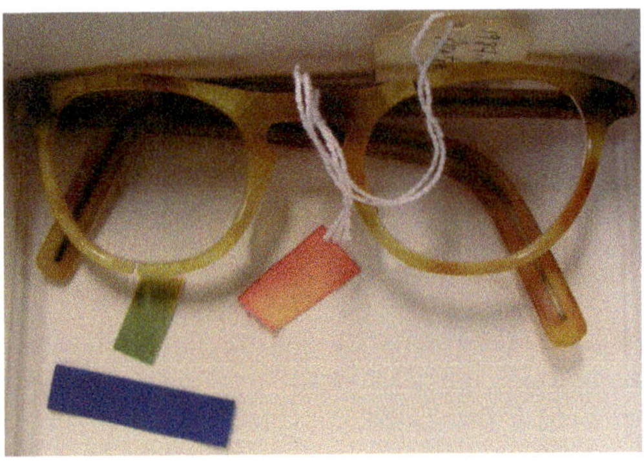

Figure 7.5 Eyeglasses monitored for three days. A-D Strip turned light green (positive reaction, compared to unused blue control strip). Cresol purple turned pink (positive reaction). Monitoring helped confirm glasses as CN, not CA. NMAH 1984.0140.383. Photographed by author, Division of Medicine and Science, National Museum of American History, Smithsonian Institution.

Cresol purple and red strips are not light sensitive so they can be left in place to monitor CN objects for years. It is important to remember that the indicators do not identify CN but rather nitric acid that may form as part of CN deterioration. So, CN that is not off-gassing, either because it is in good condition or because it has already released all the acidic vapor it can, the plastic is still CN. If reacted, once the strips are removed from the acidic environment, they will revert back to the original yellow color so any observations and recordings of the reaction should be done as soon as possible. Because they are adsorbent, indicators should not be reused. Since the use of cresol purple or red for long-term monitoring of CN may not be familiar to all collection caretakers, it is a good idea to leave directions in storage so future staff understand what the yellow indicators are, what a change from the original yellow means, why acidic vapor from degrading CN is a risk, and who to contact if a monitor changes color.

Ozone Test Sticks

Ozone Test sticks are a readily available commercial product made by Macherey-Nagel that are marketed for the detection of ozone but can get a false positive when exposed to oxidants such as chlorine (Pearlstein 2017). Because chlorine is used in the manufacture of polyvinyl chloride (PVC), as covered in Chapter 4, Ozone Test sticks were tested for their potential to monitor PVC objects.

The reactive portion of the stick starts as white and changes to beige to brown; the darker the reaction, the more oxidant present (Macherey-Nagel 2022). The degree of change for a positive reaction is determined by comparing the strip to the color scale on the packaging. The reaction typically starts along the edges since the slightly raised reactive portion has more surface area. From the date of manufacture, the shelf life of the strips is 1.5 years (Machery-Nagel n.d.). The manufacturer notes that unused strips should be stored in a dark, cool, and dry location, typically assisted by keeping them in the metal cylinder they are shipped in, and, if stored correctly, the strips can be used until the use-by-date on the packaging (Macherey-Nagel 2022).

During a series of tests on PVC objects dating from about 1985–2015, it was found that Ozone Test sticks seem to be sensitive to degrading plasticized PVC and are capable of reacting before signs of deterioration, such as weeping or discoloration, are evident. The degree of change on the monitors, however, was most pronounced in obviously degrading PVC (Coughlin 2023). When monitoring PVC, it is helpful to use Ozone Test sticks in conjunction with A-D Strips in order to get a fuller picture of any degradation products being emitted.

General Testing Protocol for Ozone Test Sticks with Small PVC Objects:

- Choose an appropriately sized glass container.
- Using clean metal tweezers, place the Ozone Test stick and A-D Strip in the bottom of the container.
- Wearing nitrile gloves, place PVC objects in the container, avoiding direct contact with reactive areas of the strips.
- Put aluminum foil over the top of the container.
- Seal aluminum foil edges with Parafilm.
- Do NOT open the container. Monitor through the container for color changes of the strips, ideally daily, for at least one week and up to a month.

IPI notes that "in the extreme case of a very small roll of film inside a very large can, an inaccurate A-D Strip reading is possible" (2022). So, the PVC object being tested should fill up as much of the container as possible. Because the indicators react to the build-up of acids and oxidants, do not open the container during monitoring or those products will dissipate. Instead, peer through the glass sides to determine if the indicators have reacted. Because it can take time for degradation products to build-up, monitor for one to two weeks. Longer monitoring is okay if desired but remember that A-D Strips can fade from light exposure and a month is about the recommended limit for testing (Hackney 2016; Townsend, Hackney, and Kearney 2023). After the designated monitoring period, compare the test strips to an unused strip and/or to the color scale provided by the manufacturers to gauge the degree of the reaction. Monitoring with Ozone Test sticks has only been carried out by the author on small, three-dimensional PVC objects in sealed containers. Protocols have not been established for larger objects or textiles.

In the testing undertaken by the author, Ozone Test sticks seem to have a sensitivity to plasticized PVC, which may point to a link between the formulation or processing of plasticized PVC and the release of oxidants as they age that these monitors react to. Rigid unplasticized PVC did not cause a reaction of either A-D Strips or Ozone Test sticks and no other type of plastic that was tested, even obviously degrading ones classified as problem plastics like polyurethane foam, caused a reaction of the Ozone Test sticks (Coughlin 2023).

Of the PVC that reacted, the ones that caused a positive reaction of both the Ozone Test sticks and the A-D Strips were more severely deteriorated, even if they looked okay. For these, FTIR analysis identified loss of plasticizer and degradation of the polymer itself (Figure 7.6). This seems to support that oxidants

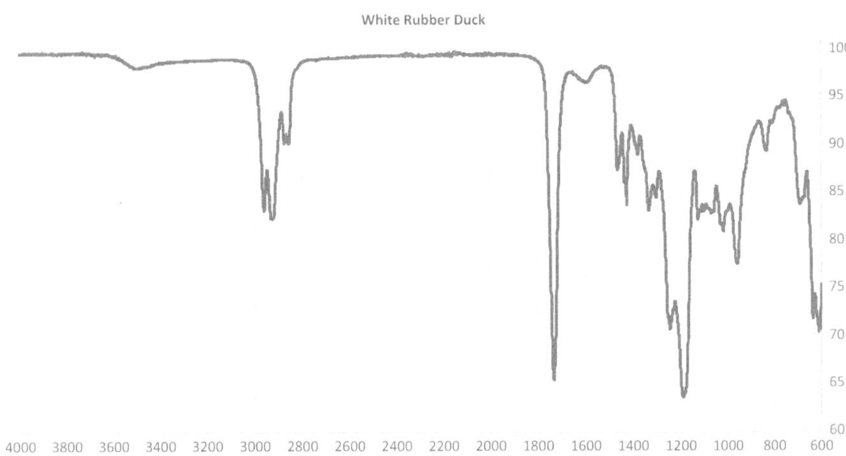

Figure 7.6 FTIR spectrum of PVC with polymer degraded and significant loss of plasticizer. Credit: Stephen Boyes, Associate Professor, Chemistry, George Washington University.

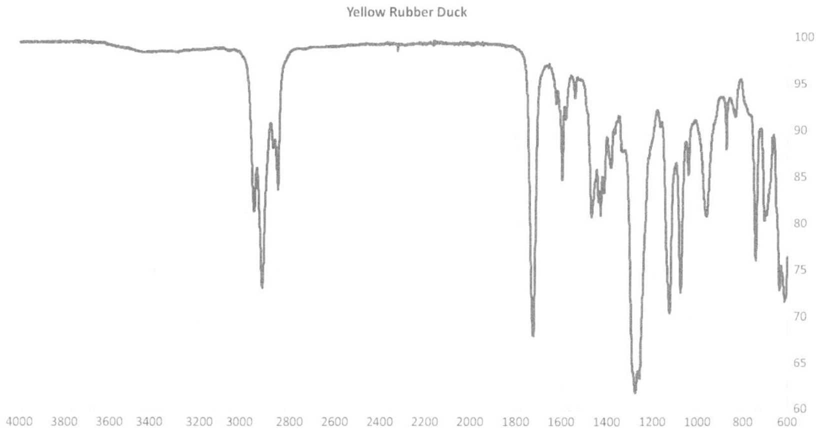

Figure 7.7 FTIR spectrum of PVC with polymer in good condition and plasticizer intact. Credit: Stephen Boyes, Associate Professor, Chemistry, George Washington University.

from PVC were being released to a degree that caused a strong reaction with the Ozone Test stick and enough acids were emitted to react with the A-D Strips, likely from the formation of hydrochloric acid and possibly from the commonly used DEHP plasticizer forming phthalic acid, as discussed in Chapter 4.

In one test, only the Ozone Test stick changed while the A-D Strips remained blue. In FTIR analysis of that sample, the plasticized PVC had relatively intact polymer and plasticizer (Figure 7.7) which seems to indicate that it was not far enough in the degradation process to release acids but was emitting products that reacted with the Ozone Test stick (Coughlin 2023). Thus, Ozone Test sticks seem to have potential as an early indicator of plasticized PVC deterioration. It is important to realize that some degree of change to the PVC has to be occurring to get a reaction with the Ozone Test sticks. This means that Ozone Test sticks cannot be used to identify PVC, since some confirmed PVC objects did not react in testing, but rather, these monitors may be able to identify when PVC is degrading even when visible signs of deterioration are not apparent.

Based on test results, Ozone Test sticks seem to have the potential to serve as an easy-to-use, affordable monitor to flag PVC degradation before noticeable signs of deterioration begin. Monitoring is recommended to be done alongside A-D Strips to provide insight into whether acids are present with the assumption that the presence of acidic vapor likely indicates the PVC is further along on its degradation path. Figure 7.8 shows the results of monthlong testing that also included copper coupons because of their sensitivity to chlorides and oxides. However, it was determined that the copper coupon did not provide additional information since it did not react while the A-D Strip and Ozone Test stick did. Though the copper coupon did not provide added insight in this example, it could help flag issues once

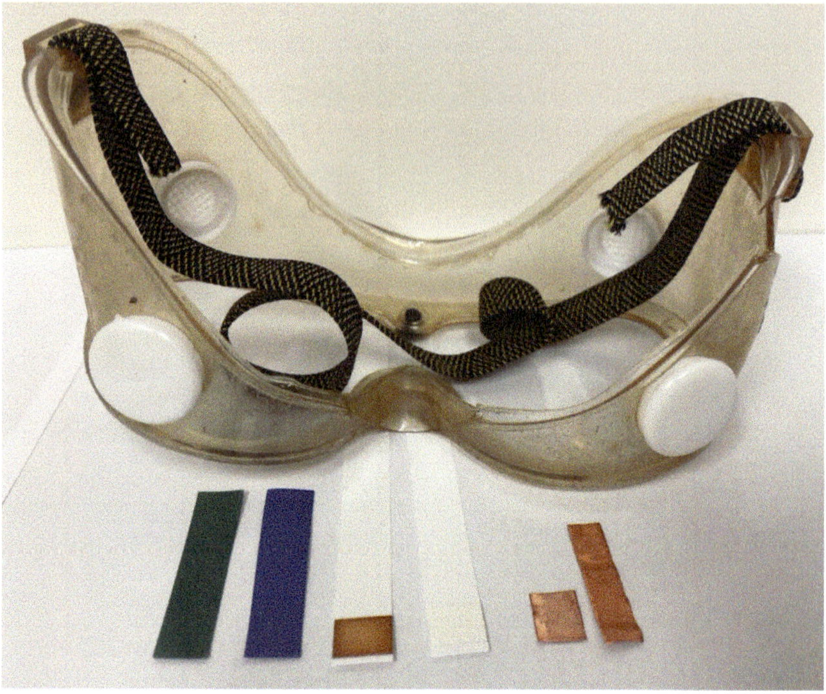

Figure 7.8 Monitoring degrading PVC goggles with A-D Strip, Ozone Test strip, and copper coupon for 30 days along with control references. Positive reactions of the A-D Strip and Ozone Test stick support FTIR results that the PVC is severely degraded. No reaction of copper coupon.

the concentrated testing period ends if left near the plastic for the longer-term, as discussed earlier in this chapter.

Key Points for Low-Cost, On-Site Monitoring of Plastics

Monitoring plastics is important to better understand their current condition and any risks they pose to nearby materials. Some monitors can be left for the long-term (metal coupons and cresol purple or red). While others provide a snapshot of the moment of testing (A-D Strips and Ozone Test sticks). Some have a wide range of sensitivities and can be used to monitor for degradation from a variety of plastics (metal coupons and A-D Strips). But some require knowledge of the plastic in question to be useful (cresol purple or red and Ozone Test sticks). Some can be easily purchased and come ready to use (A-D Strips and Ozone Test sticks). While others require preparation (metal coupons and cresol purple or red). To determine which monitors best serve your collections, it helps to have a sense of what plastics are likely present in order to know what risks they may pose, as discussed in Chapters 3 and 4. Once that is answered, decide which low-cost monitors could be deployed on-site. Then once the results are interpreted, decide if any mitigation measures are needed.

References

AIC Wiki. 2024. "Oddy Test Protocols." American Institute for Conservation. www.conservation-wiki.com/wiki/Oddy_Test_Protocols#Metropolitan_Museum_of_Art_(MMA)_Protocol

Coughlin, Mary. 2011. "Monitoring Acidic Off-Gassing of Plastics." *Conserve O Gram* 8/5. National Park Service. www.nps.gov/museum/publications/conserveogram/08-05.pdf

Coughlin, Mary. 2023. "Ozone Test Strips for PVC Plastics?" In *Plastics in Peril: Focus on Conservation of Polymeric Materials in Cultural Heritage, Proceedings of a Conference Held on Zoom, 16–19 November 2020*. Virtual, edited by Sophie Rowe, 113–123. Cambridge: University of Cambridge Museums.

Fenn, Julia. 1995. "The Cellulose Nitrate Time Bomb: Using Sulphonephthalein Indicators to Evaluate Storage Strategies." In *From Marble to Chocolate: the Conservation of Modern Sculpture Tate Gallery Conference, 18–20 September 1995*, edited by Jackie Heumann, 87–92. London: Archetype Publications.

Hackney, Stephen. 2016. "Colour Measurement of Acid-Detector Strips for the Quantification of Volatile Organic Acids in Storage Conditions." *Studies in Conservation* 61 (sup1): 55–69. https://doi.org/10.1080/00393630.2016.1140935

IPI (Image Permanence Institute). 2022. *User's Guide for A-D Strips Film Base Deterioration Monitors*, Seventh Edition. Rochester, NY: Image Permanence Institute https://s3.cad.rit.edu/ipi-assets/publications/adstrips_instructions.pdf

Macherey-Nagel. 2022. "Ozone Test." www.mn-net.com/media/pdf/ab/87/4b/Instruction-90736-Semi-quantitative-test-strips-Ozone-Test-for-Ozone-in-air.pdf

Macherey-Nagel. n.d. "Semi-Quantitative Test Strips Ozone Test, for Ozone in Air." Macherey-Nagel. Accessed May 8, 2024. www.mn-net.com/us/semi-quantitative-test-strips-ozone-test-for-ozone-in-air-90736

Nicholson, Catherine and Elissa O'Loughlin. 1996. "The Use of A-D Strips for Screening Conservation and Exhibit Materials". *The Book and Paper Group Annual* 15. American Institute for Conservation. https://cool.culturalheritage.org/coolaic/sg/bpg/annual/v15/bp15-11.html

Odegaard, Nancy, Scott Carroll, and Werner S. Zimmt. 2015. *Material Characterization Tests for Objects of Art*, Second Edition. London: Archetype Books.

Pearlstein, Ellen. 2017. "Teaching Sustainable Collection Care." *Journal of the American Institute for Conservation* 56 (2): 113–125. https://doi.org/10.1080/01971360.2016.1217726

Stephens, Catherine H., Eric M. Breitung, Corina Geiculescu, and Thomas Strange. 2022. "Determination of Nitrile Gloves Appropriate for Use When Dry Handling Art." *Journal of the American Institute for Conservation* 61 (3): 162–171. https://doi.org/10.1080/01971360.2021.1913543

Townsend, Joyce H., Stephen Hackney, and Mark Kearney. 2023. "A Critical Assessment of A-D Strips® Used in Degradation Studies of Artworks Made from Cellulose Acetate." *Studies in Conservation* 68 (1): 9–17. https://doi.org/10.1080/00393630.2021.1996091

8 Plastics Collection Management Part 1

Processes and Procedures

Up until this point, the book has focused primarily on preventive conservation – figuring out which plastics are present, how environmental factors and material choices affect them, and how to monitor and mitigate issues. This chapter marks the switch to collection management decisions as they relate to plastics. Collection management content is broken into two parts: practical actions involving processes and procedures followed by a chapter on nuanced discussions and decision-making. This first chapter on collection management of plastics covers:

- Handling
- Documentation
- Databases
- Labeling

Handling

When handling plastics, assume fragility and the potential for health risks. Remember, sometimes a plastic can look fine but actually be degraded such as plasticizer loss causing embrittlement that can lead to cracks and breakage and bring with it health concerns when handled. Or, as the case with textiles, they may no longer be elastic and could sag and deform if unevenly supported or essentially be locked into place (Sidebar 8.1).

Sidebar 8.1 Handling Human-Made Fiber-Based Textiles

Kris Cnossen

Textiles are inherently flexible objects which need support. This support can be provided through a mannequin or display mount, specialized storage system, or the use of a sling during handling. Slings can be made from clean Tyvek or undyed muslin. To protect against creases, flat textiles may be

DOI: 10.4324/9781003489658-8

rolled or folded and padded with archival tissue rolls or polyester batting-filled stockinette tubes placed in the folds. Three-dimensional textiles require padding during flat storage, while hanging storage is often suitable for garments in good condition. These supports are vital for a textile made from plastic because they allow the textile to maintain a legible shape as it ages. Creases in human-made fiber-based textiles are difficult to reduce, making padding out to lessen the occurrence of creases important. Silicone paper can be a good barrier material for plasticized plastics, especially if they need to be padded out or if they touch another part of the textile object in storage, such as a jacket hanging on a storage hanger.

When it comes to handling plastics in collections some questions routinely come up: Are there health and safety concerns? Are cotton gloves okay? Can I reuse nitrile gloves after handling plastics?

As covered in Chapters 2 and 4, plastics can poise health concerns both from handling and inhalation of emitted volatile organic compounds (VOCs). If a plastic has an odor, whether it is new and VOCs are being emitted or is aged and degradation products like acids are off-gassing, examining plastics under ventilation is a good idea. Since the components that can exude from them, like plasticizers, can be a health concern, always handle plastics with gloves. Powder-free nitrile gloves have become the standard default for handling collections but they should not be worn to handle multiple plastics, especially if they are degrading, because of the risk of transferring degradation products between plastics. Additionally, the chemicals being exuded may be able to breakthrough the glove barrier so if liquid exudate gets on gloves, remove immediately, dispose as hazardous waste, wash hands, and put on a new pair of nitrile gloves (Ledoux et al 2023).

Other types of gloves may pose risks to the wearer, object, or both. Latex gloves can cause an allergic reaction in some people with increased exposure resulting in higher sensitivity (Stephens et al 2022). Additionally, as is the case with some nitrile gloves discussed in Chapter 7, latex gloves can contain sulfur which can corrode some metals and, therefore, are not recommended for collection use (Hatchfield 2002).

Vinyl gloves also come with a risk of allergic reaction but this is less than with latex (Barker 2010). Because vinyl gloves are made of plasticized polyvinyl chloride (PVC), they have the associated risk of phthalates migrating out which poses a health risk to the wearer and can contaminate other collections and are, therefore, not recommended (Stephens et al 2022).

Cotton gloves may be appealing because they can be washed and reused, which supports sustainability goals. However, they are not frequently washed meaning that dirt and oils can build up and transfer between objects during handling (Stephens et al 2022). The porous nature of cotton works both ways to put both the wearer and the object at risk since it allows sweat and oils from hands to wick through the cotton to objects and any substances on surfaces, such as weeping plasticizers, to work their way through the gloves and be absorbed by the wearer. Thus, cotton gloves inadequately protect both the wearer and the object. Cotton gloves are also often ill-fitting so can get caught on surfaces, are relatively thick which reduces tactile feedback, and can leave cotton lint behind (Barker 2010). All of these factors combine to make cotton gloves an outdated and ill-advised choice for handling collections, especially plastics.

Do not assume current or future staff know handling risks for plastics! It is important to note in database entries for plastics the requirement to wear powder-free nitrile gloves and to dispose of them as hazardous waste after handling a degraded plastic.

Documentation

Documentation is a key part of the work of collection stewards and takes on extra importance with plastics because good written and photographic documentation can help determine if changes have occurred over time and to what extent, which can then inform decisions for care. In some cases, documentation may play a role in insurance claims so that insurers can try to determine if the causes of the claims are tied to anticipated aging of the plastic (Wessel 2008). Documentation can also inform decisions of whether or not to deaccession. If a plastic is so degraded that the decision is made to deaccession (Chapter 9), then documentation is all that will remain of the plastic in the collection. If a work is collected from a living artist/maker/creator, then documenting why they chose the plastic(s) they did and their views on how they may change in appearance and functionality over time is an insightful and important part of documentation that can help inform future decisions, as discussed in Sidebars 8.2 and 9.2.

Sidebar 8.2 Artist Interview

Steven O'Banion and Samantha Owens

Acknowledgment: The methods and ideologies presented here have grown out of the VoCA Artist Interview Workshops, which were developed by VoCA Board members Richard Cándida Smith and Jill Sterrett, then refined over the years by a series of workshop leaders.

When considering an artist interview, it is important to ask oneself, "Why?" An interview is a commitment of time and attention for both the interviewer and narrator. If one has just a few questions, often research, analysis, and/or a written communication are sufficient. Interviews are usually better suited when one hopes to garner a deeper understanding of an artist's work by elucidating their intentions, practices, and memories. An interview can clarify why an artist chose to incorporate plastics into their work, which will allow one to tailor a preservation strategy for the artwork that considers both the materials used and the artwork's intended meaning.

A keen sense of self-awareness is invaluable when preparing for an interview. The conversation often is entered into with specific objectives in mind. The interviewer brings with them, whether consciously or not, predispositions and biases. There can be unbalanced power dynamics at play, age gaps, gender roles, and contrasting belief systems. By acknowledging these types of factors, the interviewer can better understand the narrator's position and adapt their interview style accordingly.

When deciding who will participate in an interview, be aware that too many voices can result in a dialogue that is unfocused or hard to follow. Colleagues will sometimes request to join in and ask questions of an artist. Similarly, an artist's relations or gallerist may want to provide emotional support or feel the need to fill in missed details. While this can be managed successfully by skilled oral historians, generally speaking, having two to three total participants will lead to a more coherent and meaningful transcript.

Thorough preparation can demonstrate to a narrator the interviewer's sincerity and help to build trust. It is the responsibility of the interviewer to research an artist's biography and body of work, as well as familiarize themselves with previous interviews. Further, practicing with the equipment one will use to record the interview will prevent unnecessary fumbling in the narrator's presence and prevent the loss of a recording. Participants in an interview own their words. As such, the interviewer should secure a signed legal release form that outlines acceptable uses of any recordings or transcripts, ideally when the interview is completed. If prepared, one will be able to move beyond potential canned narratives by following an artist's references and asking informed, novel questions.

The location of an interview can profoundly affect the outcome. Being in front of an artist's work or at their studio can help to bring back memories. When this isn't possible, images of artworks can be helpful tools. Moreover, no one's memory is infallible, and it should not be expected to be. Evoking all the senses when posing questions, and inquiring about processes, rather than facts, can often lead to deeper and more reliable information.

There is a misconception that a well-planned series of questions will, by default, lead to a smooth interview. However, a rigid sequence of questions can be restrictive and disrupt organic revelations. An alternative approach is to identify a few topics of interest and be prepared to bring them up as they fit into the discussion. A successful interview will find a natural back-and-forth cadence. It is a dialogue, not a monologue or soliloquy. After asking a question, remember to pause. Habitually, one will want to speak to fill a silence. However, the narrator may be taking time to think. Frequently after a long, considered pause, a narrator produces their most profound statements.

An interview captures the content of a conversation among a specific combination of people at a discrete point in time. Though, by establishing a lasting relationship with an artist, one stands to build trust that can lead to a series of interviews over an extended period. Following up with the artist by remembering to send them copies of recordings or transcripts, if promised, and keeping them informed about projects related to their interviews, is essential to maintaining a good rapport. The deep understanding of an artist's work will allow one to make informed decisions regarding its storage, preservation, and exhibition.

Documentation of an object begins by looking, not touching. After looking over the overall form, examine the object in a logical manner such as front to back or side to side. Then record your observations in writing and in photographs. The breadth and depth of the information recorded will depend on whether examination is for a condition survey, which likely aims to capture content in a uniform and relatively quick manner for a lot of objects, for an incoming or outgoing loan, or for a conservation treatment report.

The method of entry for documentation is typically into an electronic database but in some cases may still take paper form. In either format, the degree to which responses to categories are open-ended or streamlined, such as using a pre-set list of materials and condition descriptors, affects the time it takes to input information and how that information can be utilized later.

Documentation is clearly tied to database entry, discussed below, and both benefit from streamlined approaches for the type of information that gets recorded. The general sections for documentation of plastics should include the following:

- **Marks** (catalog number, maker's marks, Resin Identification Code)
- **Materials**
- **Material identification method** (scientific analysis, marks, appearance, history, etc.)
- **Form** (rigid, flexible, foam, film, fiber)
- **Measurements**
- **Weight**
- **Color**
- **History** (date made and date entered museum)
- **Manufacture**
- **Condition**
- **Conservation priority and needs** (including rehousing)
- **Curatorial significance**
- **Handling recommendations**
- **Environmental recommendations**
- **Housing** (current and recommended)
- **Photographs**

Identifying marks from the time of manufacture and numbers added from accessioning should be noted to ensure that the database entry matches the object. If a more concrete identification of the plastic is possible, note how it was determined. If scientific analysis was undertaken, what tests or instruments were used? If identification is based on appearance or other physical characteristics or on history of use or other curatorial information, note that. If an identification is likely but not confirmed, it is still helpful to include but with caveats like "maybe," "possibly," or "likely" or using the convention of adding (?) after the name. For example, if a plastic is likely PVC but there is some uncertainty, it could be noted as "possibly PVC" or "PVC(?)." Or have a pre-set ranking of confidence in the plastic identification as low, medium, or high.

If more specific identification is not possible, it is helpful to note the original form in general terms like: rigid, flexible, foam, film, or fiber. This will help determine changes over time as well as guide storage and display decisions, such as using silicone coated Mylar/Melinex or silicone release paper at points of contact for flexible plasticized plastics and ensuring adequate support in case flexibility lessens over time.

Measurements (height, width, depth) are important to help confirm that you have the correct object and changes to measurements over time can be used to indicate deterioration since some plastics will distort as they degrade and lose mass. If weeping material from the polymer matrix sublimates from the surface, mass is lost which can be noted by weighing, as discussed in Sidebar 4.2. Additionally, for plastics that absorb moisture in high relative humidity (RH), such as Nylon which can take on 10% of its weight in water, an increase in weight would point to the RH being too high (van Oosten 2022).

Color is important to note in the description as a quick way to make sure you have the right object as well as to determine if the color has changed over time (Sidebar 8.3). Though determining if color has altered is usually carried out by comparing the object to images, it is important to note the color in writing in the database as well in case the image files are corrupted or lost. Recording color in writing can be subjective though – my dark blue may not be your dark blue or you might enter a more descriptive shade such as Navy blue or Royal blue. So, as with naming plastics (Sidebar 8.5), it is important to establish naming protocols for color, such as agreeing on a color palette to use as a reference (Sidebar 8.3). If this is pursued, ensure that those entering and using the database have access to the color reference system and understand the rationale for its use. If a plastic is transparent or translucent, note that as well since changes in opacity can occur as part of aging.

Sidebar 8.3 Identifying and Tracking Color Change

Christopher Mazza

In 2016, the Costume Institute acquired a newly made Chanel suit (2016.379a–c) with a 3D printed polyamide surface (see Sidebar 2.2). Consistent with other 3D printed objects in the collection, yellowing was observed after only a few years in storage. This color change was confirmed by contrasting newly taken images of the suit to acquisition images that contained a white balance card.

To improve identification and track future color change, two steps were taken. First, to serve as a quick reference, the polyamide surface was matched to an equivalent Pantone color. When using Pantones, a consistent light source provides the best conditions for making both a color match and for future comparisons. Alternatively, a closely matched fabric or paint swatch can serve as a color reference and, if archival, can be stored alongside the object if kept in low light conditions.

Second, to quantitatively track the suit's discoloration, a spectrophotometer reading was taken to establish a baseline of L*a*b* color values and ΔE from which to evaluate future readings, as discussed in Chapter 5. While less precise, Photoshop can also be used to determine L*a*b* and ΔE. Given the potential unevenness of yellowing on 3D printed surfaces, it's crucial to select and document a consistent location for all readings.

Information about the history of the plastic including when it entered the collection and when it was likely made, even if that is a date range or generically listed as a decade, can be helpful in narrowing down possible plastics for identification (Figure 3.6). Insights into its manufacture such as the country of origin whether it was mass produced, or if it was mixed by an individual creator can all help further understanding of the plastic.

Of course, the current condition is essential for documentation and should be noted in detail and supplemented with photographs. Having an overall ranking of the plastic helps give a quick snapshot of its state at the time of documentation but this is also subjective, what I may consider good condition, you may note as fair. So, deciding on terms and what they mean is helpful. This could take the form of numerical ranking and/or descriptors such as:

- **5 Excellent**: pristine condition
- **4 Good**: minor surface issues
- **3 Fair**: some structural issues or more extreme surface issues
- **2 Poor**: structural issues or severe surface issues
- **1 Very Poor**: severe structural and surface issues

Tables 4.1 and 8.1 include frequently used terms for visible signs of plastic degradation that could be incorporated into a pre-set list of condition issues for use with plastic collections, again with the goal of saving time and streamlining results. Having a visual condition atlas can further help with the effort to document condition as described in Sidebar 8.5. Other visual references to help describe plastic degradation can be found in Appendix 3 of Yvonne Shashoua's *Conservation of Plastics* book (2008), the "Damage Atlas" from the Preservation of Plastic

ARTefacts (POPART) project (n.d.), as well as more specialized resources like the "Visual Decay Guide" for film (Image Permanence Institute (IPI) 2024). Though not focused on plastic, the Australian Institute for the Conservation of Cultural Materials (AICCM) website has a visual glossary of condition issues with images and descriptions that may be helpful (n.d.).

Recommendations for care are not limited to interventive conservation treatments but if that is decided, it is helpful for planning to estimate time and/or cost, if that is in the skillset of the person undertaking documentation. Preventive measures like the need for better housing in storage or the benefits of being placed in a specialized environment should be indicated, again with estimates for time and/or costs, if possible. To make it easier on the surveyor and those searching the database later, have pre-set options for preventive care such as:

- **Current housing** (have pre-set list of common storage methods at your institution)
- **Housing recommendations** (may have pre-set options such as: "replace with archival corrugated board," "remove from polyethylene bag," "add scavenger," "increase ventilation," and "acceptable current housing")
- **Monitor** (Chapter 7)
  - Metal coupons (silver, copper, lead)
  - A-D Strips
  - Cresol purple or cresol red indicators
  - Ozone Test sticks
- **Light levels**:
  - Storage: standard storage levels or dark
  - Exhibition: usually about 50 Lux or 5 foot-candles (Chapter 5). Helpful to provide duration and rotation guidance.
- **Low temperature storage**: cool, cold, frozen (ranges in Chapter 5)
- **Anoxic storage** (Chapter 5)
- **RH ranges for plastic type** (Chapter 5):
  - Standard RH: 45–55% with allowable drift of +/–5%
  - Low RH: 20–30% (cellulose acetate (CA), cellulose nitrate (CN), and polyurethane (PUR) esters)
  - Higher RH: 55–65% (casein formaldehyde (CF) and urea formaldehyde (UF))

For most collections, condition alone is not the sole driver for whether a plastic will get put into a specialized environment or receive attention from a conservator. Curatorial priority and how critical an object is to the museum's mission is often taken into account when prioritizing care. For some museums, this can take the form of tiered ranking and is helpful to capture in the object record.

Remember that the condition as well as the plastic type can inform handling and environmental decisions as well as guide choices for storage and exhibition materials, as covered in Chapters 5 and 6. These recommendations should be included in database entries for each plastic.

An important element of documentation is photography. This should include overall images of all sides as well as details of any important elements and emerging condition issues. In the overall images, include an identification bar with the accession/catalog number, date the image was taken, a white balance card, and a color bar to remove some of the subjectivity later if determining if color has changed (Sidebar 8.3). To help counter color interpretation issues further, make sure that the camera, computer screen, and printer are color-balanced so that comparisons between them are uniform. Using a tethered capture software to take the photos can be a relatively affordable way of taking consistent color-balanced images. If ultraviolet (UV) examination picks up any issues, consider taking photos and inputting them into the file (Figures 10.1 and 10.2). It is helpful to note what UV lamp was used and at what wavelength so that future examinations can replicate the setup for comparison.

Because there is a lot of information readily available about image file choice and associated data, it will not be covered in-depth here. That said, remember that metadata for digital files is important for searching, retrieval, and readability of the images and can help prevent dissociation of the files. The format the photo file is saved as also matters. JPEG images are smaller but get compressed and can lose data each time they are opened so is not a good choicc for collection management purposes. For record keeping, TIFF is best but the files are large which may be difficult to utilize if storage space is limited. And, of course, backup electronic media and keep up with updates and changes to technology so electronic records can be accessed in the future.

Databases

Without an organized database to input documentation, the management of collections is not only more difficult but can stray into dissociation that can result in loss of objects and/or data (Cato and Waller 2023). While most institutions have standardized information to include in records inputted into a collection management system, the level of description and other required information varies. For plastics, this potential lack of information can have consequences for care because if you do not know what plastics you have, you do not know their environmental needs and what risks they may pose to other materials.

So, what information is helpful to include about plastics in a collection database? Sidebars 8.4 and 8.5 identify needs and approaches to incorporate so that the content is more helpful and better serves the needs of the collection today as well as anticipating how the database may be mined for information in the future. A powerful step in strengthening care through database content is streamlining terminology so that as multiple staff members or volunteers input information, it is more uniform and, therefore, searchable (Sidebar 8.5). The Design Museum Gent developed two thesauri for plastics in their project "Know, Name and Assess your Plastics." One thesaurus focuses on plastics as materials and the other covers processes and techniques of manufacture. The project notes that "Both thesauri can be used separately, but complement each other when combined" (Design Museum Gent 2022). The thesauri fill a gap since

> Most institutions lack specific terms in their registration systems to categorise these materials, more specific than the term 'plastic'. An efficient thesaurus would enable museums to enter their findings more precisely into their collection registration system and improve the searchability of their database.
>
> (Design Museum Gent 2022)

Much of the difficulty for many museums in reaching this organizing goal is the varying knowledge and confidence levels in plastics identification of those entering data. Since many institutions lack protocols, people utilize their own knowledge so that someone with a background in chemistry or more experience identifying plastics, even without scientific analysis, may enter the polymer family name while someone else may use a product name or just list "plastic." Even if it is agreed to use the polymer name, the spelling of those can vary, as with PVC which in different publications in the museum field shows up as "polyvinyl chloride" or "poly(vinyl chloride)." Searching for one or the other in most databases or searching for the abbreviation PVC may not pick up results for the other spellings thus making the results incomplete. For most collection databases, people just input "plastic" as the material regardless of whether it is flexible or rigid, or a foam, sheet, film, or textile. When someone later searches the database for "plastic," a torrent of results is likely not helpful just as an entry that is too specific or obscure and would likely not be entered as a search term is equally unhelpful. By making the options for input uniform, those adding to the database and those searching it later will achieve better results. If more concrete identification is used, it is helpful to note how that determination was made: scientific analysis, physical characteristics, or curatorial information such as its history of use. The "Know, Name and Assess your Plastics" resource is a good starting point for guidance on entering plastic data but it will be limiting for some collections since it does not include human-made fibers so terminology for plastic-based textiles will have to be sorted out by each institution (Design Museum Gent 2022).

If collection staff are unable to follow the nomenclature recommendations laid out in the "Know, Name and Assess your Plastics" thesauri, which does require a strong familiarization with plastics and confidence in identification, there is an approach that is a compromise between naming a polymer and using "plastic" as a catchall term. For those seeking an option that is more specific than "plastic" yet general enough for broader application, consider establishing guidelines that note the plastic form. That way, changes in characteristics from the original database entry, like a flexible plastic that is now rigid, provide condition insight. Also, noting something as a foam, which is typically a vulnerable form, will enable searches for foams to be conducted so conditions can be periodically evaluated. Of course, if more specific identification can be added, perhaps by applying the guidance for identification in Chapter 3, that should be included in the database.

Categorizing plastics under the three main types from Chapter 2 could be helpful:

- Thermoset
- Thermoplastic
- Elastomer

Or using more general terms may be a more accessible approach:

- Rigid
- Flexible
- Foam
- Film
- Fiber (noting if that fiber is meant to be elastic)

Whichever route is decided, it is vital to note that decision so that those entering data and those searching it later follow the same protocols and procedures in order to make the database as functional a tool as possible.

Sidebar 8.4 Designing Plastics Surveys for Successful Follow-Up

Jessica Walthew, Kate Wight Tyler, and Sarah Barack

Surveys offer efficient means to gather information about collections and help to smooth uneven data across object types or media. An organized survey with a clearly articulated scope provides vital insight that captures both verified information and areas needing research, providing the foundation for preservation planning and grant submissions.

In 2011–2013 project conservators surveyed Cooper Hewitt's plastics collections, examining 1,553 objects over 18 months. They assessed objects' condition and storage housing, identified basic categories (i.e., hard or flexible plastic), and made recommendations for interventive treatment and preventive conservation. The survey used visual and tactile methods to categorize plastics according to characteristics (e.g., hard versus soft) including manufacturing data when available. Analytical identification was restricted to a small subset of objects (about 20) analyzed by Fourier transform infrared (FTIR) and Raman spectroscopy by external conservation scientists.

A tiered system was developed to triage the works into categories for both housing and condition, showing that while 81% of the collection was in good or fair condition, 52% of the collection had poor to unacceptable housing.

The survey was carried out in Microsoft Access with controlled vocabulary fields whenever possible; results were then migrated into The Museum

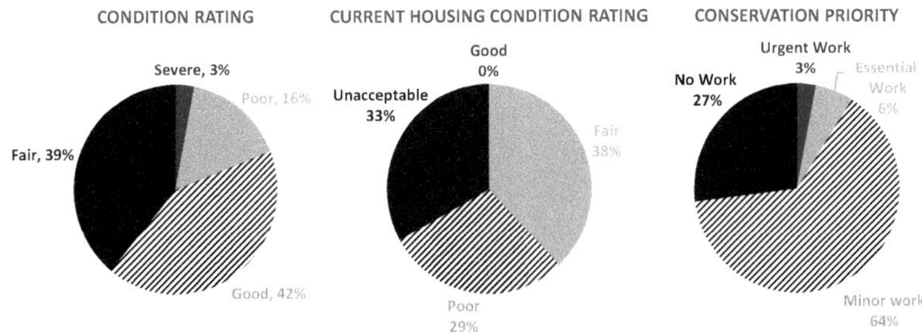

Figure 8.1 Condition, Current Housing, and Conservation Priority pie charts showing proportions of the collection needing attention for housing and treatment, and the prioritization recommendations. Credit: Kate Wight Tyler.

System using text fields for easy searchability. Importantly, hourly estimates for conservation treatment and rehousing were provided at an object level which enabled better planning for subsequent projects. A total of 1,328 hours of urgent treatment were recommended; 60% of objects lacked labels, with 172 hours estimated to remedy this!

The final reports from the project conservator included top priorities for follow up:

- Immediate rehousing needs were addressed during the survey, with most plastics relocated to a new "Plastics room" to facilitate monitoring. A low temperature storage vault (40°F(4°C) and 40% RH) was built but had to be taken offline due to the system's inability to maintain set points.
- Semi-synthetics (i.e., CA and CN) are uncommon in the collection; examples were segregated for periodic monitoring with A-D Strips, and housed with acid adsorbing microchamber board. By contrast, elastomers (both natural and synthetic) in poor condition were found throughout the product design collection as shock absorbers and other components. A pilot anoxic project is ongoing but requires further research.
- Furniture with PUR foam upholstery or PVC elements were identified as needing urgent treatment. Fellowship projects on both topics have enabled further analysis and treatment of several works in severe condition, as seen in Figure 4.8.
- Continuing work to better identify and monitor plastics is now supported by analytical capacity (colorimetry and FTIR instrumentation), a Plastic Identification Toolkit (from van Aubel van Rooijen), and the purchase of reference materials.

Since 2013 the initial survey has guided work on tailored projects at both an object-level and collection-level. New priority research topics have emerged on 3D printed plastics and bioplastics, showing that while initial surveys set the stage for collection management and conservation planning, surveying and monitoring plastics collections is an ongoing job.

Sidebar 8.5 Tools for Surveying Plastic Collections: Monitoring Aging Phenomena

Elena Gómez-Sánchez, Cristian Mazzon, and Till Krieg

Knowing what each polymeric material is, and, more specifically, where the most vulnerable of them are, is probably the single task that can have the largest impact on the wellbeing of plastics and rubbers in collections. The most reliable way of doing this is via non-invasive analytical techniques such as Attenuated Total Reflectance Fourier Transform Infrared Spectroscopy (ATR-FTIR). However, this option is not always available and does not work with all polymeric materials, such as many rubbers (Krieg, Mazzon, and Gómez-Sánchez 2021).

In the absence of reliable material identification, as is the case in most museums, describing and monitoring aging phenomena gives essential clues on conservation needs and how to address them. On the one hand, degradation characteristics of certain plastics can help identify them. On the other hand, if a plastic is changing rapidly (for example, if a second inspection after 12 or 24 months reveals its condition has noticeably worsened), then it should become a priority to address. With training and experience, a workflow can be put in place that will provide necessary information to improve the survival chances of these materials, or at least help increase their lifespan.

A systematic survey of the collection, where a number of objects is inspected and the results documented, can thus be very helpful when looking to identify where the conservation needs of a collection are. Surveys can be designed to answer different questions. Depending on the questions to be addressed, the sampling of objects will occur in different ways, which determine the informative value of the survey results.

Surveys aiming at inspecting and documenting aging phenomena need to rely on:

(a) objective descriptions of aging phenomena
(b) a means of documenting the appearing phenomena, so that historical information can be consulted and, ideally, compared
(c) a way to "quantify" the extent or spread of the observed phenomena

(a) Objective Descriptions: The Importance of an Atlas of Aging Phenomena

A useful tool for the first purpose is a catalog, with pictures, of all the possible aging phenomena that may appear. This allows recognizing, assessing, and documenting damage reliably and efficiently. The data need to be objective and consistent: the same kind of damage should always be labeled the same way. This implies that all staff carrying out the survey, today and in the future, have a similar picture in mind for any given term. This reduces variability in the data and increases consistency.

For this, each term describing an aging phenomenon needs to be described in a way that makes sure that it is as clear in what it includes as what it does not include. There should be no overlap and no gaps between descriptions. The name "atlas" describes quite well what is needed: as in an atlas, all the areas of the globe surface are given a single name, and no place belongs to two different areas, but is univocally assigned to one, and one only. Also, by covering the whole surface with terms, everything that can appear can be named, and no new names are needed.

Currently, there is no established set of aging phenomena of plastics with definitions over which there is worldwide consensus, but the literature offers several works which can be adopted and, if necessary, adapted (see Krieg, Mazzon, and Gómez-Sánchez 2021 for a recent review). There is good reason to have a set of terms with which staff feel confident (Taylor 2014), so they will be able to apply it with success and train new staff.

An example of a visual condition atlas that fulfills the criteria mentioned above (Krieg, Mazzon, and Gómez-Sánchez 2021) is summarized in Table 8.1. The 39 visual signs are organized in four main categories: **deposits**, **chromaticity and transparency**, **deformations,** and **loss of integrity**. Additionally, there are two further categories: five olfactory signs under the category **smell**, and **other/miscellaneous** signs which may be important to document. The main categories are mutually exclusive but, of course, more than one aging or degradation phenomena can appear simultaneously.

The category **deposits** flags where some material is noticeably covering the surface of the plastic. **Chromaticity/transparency** marks when the color or transparency characteristics of the surface have changed. **Deformations** denotes changes in the form or shape of the object, without loss of integrity. Integrity is understood here as "the quality of being whole and complete." Where integrity is challenged, then the category **loss of integrity** comes in handy. This category is further divided in two groups: (a) those where loss of material is usually involved (e.g., abrasion or crumbling), and (b) those where the observed phenomenon does not necessarily involve loss

Table 8.1 Summary of the Six Main Categories of Degradation Phenomena and the Respective Visual Signs

1. Deposits	*2. Chromaticity/ Transparency*	*4. Loss of Integrity*	*5. Smell*
a) Solid deposits a. Blooming (crystalline) b. Blooming (other) c. Dried fluid d. Dirt e. Dust f. Pox g. Stain	*a) Color change* a. Yellowing b. Fading c. Discoloration *b) Loss of transparency*	*a) Usually involving loss of material* a. Abrasion b. Chip c. Crumbling d. Flaking e. Scratch f. Loss of material (other)	a) Acrid b) Camphor c) Naphthalene d) Rubber e) Vinegar
b) Liquid deposits a. Smeary b. Sweating	*3. Deformation* a) Blistering b) Channeling c) Dent d) Fold e) Shrinkage f) Softening g) Warping	*b) Not usually or necessarily involving loss of material* a. Brittleness b. Break c. Crack d. Craquelure e. Crazing f. Hardening g. Loose h. Peeling	*6. Other/ Miscellaneous* a) Adhesive tape b) Biological attack c) Corrosion (metal part) d) Overpaint e) Sticky

of material (e.g., a crack). A further set of **other/miscellaneous** terms that are not intrinsic to the aging of the plastic itself but would be helpful to know, is included. For example, metal corrosion in the vicinity of the plastic may signal the presence of a malignant plastic in advanced state of degradation emitting corrosive, acidic gases.

(b) Documenting Aging Phenomena: Database

A convenient way to document and keep track of the information gathered about plastic condition is by creating a database. An example of an IT tool developed for this purpose is based on Microsoft Access software (Krieg, Mazzon, and Gómez-Sánchez 2021). The developed database is organized in information layers, going from general to more specific:

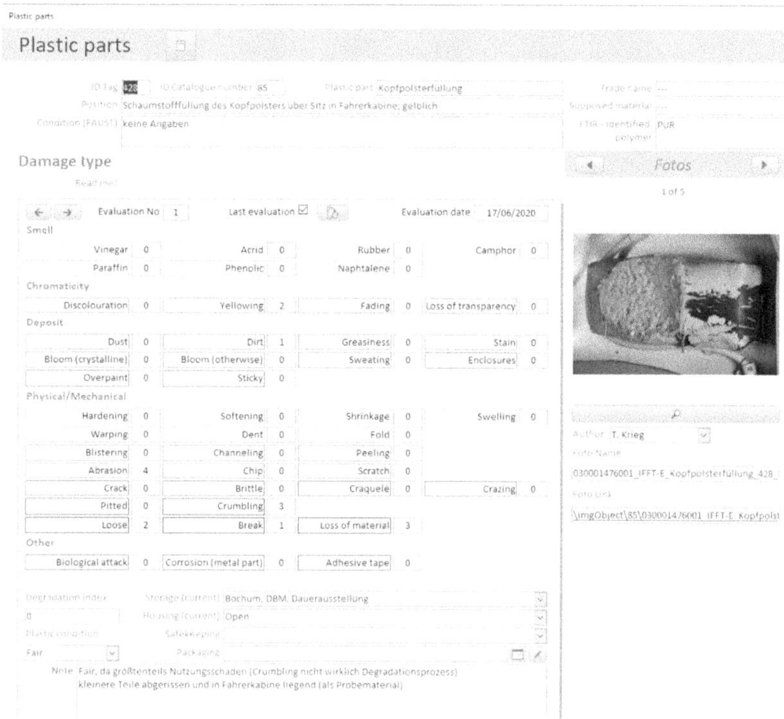

Figure 8.2 Database entry containing the categories and respective signs of degradation that can be assigned to each individual plastic part of the object ("Plastic parts").

1. Survey Catalog: list of objects inspected in the survey.
2. Object Information: metadata about the object, fields useful for cross-referencing with other museum databases, and a list of plastic parts to enable documentation of their individual state of preservation, which may be very different (Figure 8.2).
3. Plastic Parts: extent of aging phenomena documented for each part (Table 8.2).

A further layer, sampling, not presented here, gathers information on the results of analyses performed either directly on the plastic part (non-invasive analysis) or on samples.

(c) Quantifying the Extent of Aging Phenomena

It is helpful to quantify condition, typically using a scale where numbers correspond to the degree of degradation of the object as a whole (Keneghan 2001) or to the plastic part (Bell et al 2022; Bacci et al 2012). Going a step further, and since each plastic part can show different aging phenomena, it seems necessary to give each of the phenomena a separate rating (Krieg, Mazzon, and Gómez-Sánchez 2021). Whether the numbers are being applied to the object as a whole, to a given plastic part, or to a specific aging phenomenon in a given plastic part, the meaning of the number rating needs to be defined before the survey begins. An example of ranking the impact of degradation on a 0–4 scale could be from barely having an effect (0) to being unacceptable (4), as seen in Figure 8.2 and Table 8.2.

Table 8.2 Overview of the Different Numerical Scales that Are Used in the Literature to Rate the Extent of Degradation

Per Plastic Part (Bacci et al 2012)	*Per Degradation Phenomenon (Krieg, Mazzon, Gómez-Sánchez 2021)*	*Per Plastic Part (Bell et al 2022)*
0: No visible changes	0: Barely	1: Good: visible degradation slight and/or limited
1: Just noticeable changes	1: Minor	2: Fair: visible degradation perceptible and/or moderate
	2: Distinct	
2: Evident changes	3: Critical	3: Poor: obvious and/or considerable
3: Very strong changes	4: Unacceptable	4: Unacceptable/severe/ unstable: visible degradation important and/or generalized

Sidebar 8.6 Collections Survey Planning at the 9/11 Memorial & Museum

Kate Fugett and Kerith Koss Schrager

In 2023, 9/11 Memorial & Museum conservation staff completed a pilot survey to assess the condition of a representative sample of plastic-containing materials in the museum's collection. The preliminary survey, which involved a review of the collection management database and a hands-on survey of 112 objects, helped identify staffing, time, supplies, equipment, and important preparatory steps needed to streamline a future, comprehensive survey of the entire plastic-containing collection. This process also highlighted issues, specific to the 9/11 Memorial & Museum and other trauma and rapid response collections, which should be considered in the design of these surveys.

The Museum's collection includes 22,000 physical artifacts to support the institutional mission which

> remembers and honors the 2,983 people killed in the horrific attacks of September 11, 2001 and February 26, 1993, as well as those who risked their lives to save others and all who demonstrated extraordinary compassion in the aftermath of the attacks.

As a cross-section of early 21st-century life, it is not surprising that the salvaged remains of the events of 9/11 contain a significant variety of plastics. Conservators had observed characteristic plastic deterioration on individual objects over time; however, due to the extraordinary circumstances that led to the formation of the museum and collection, the condition of most of the plastic-containing artifacts was unknown almost 25 years after the events of 9/11.

Accurate documentation and material identification was a challenge from the inception of this collection. Under the intense conditions of a rescue and recovery operation, as well as an ongoing criminal investigation, 1.8 million tons of material was removed from Ground Zero, the site of the collapsed Twin Towers, and transported to Fresh Kills Landfill on nearby Staten Island. Objects selected by cultural heritage workers granted access to the site ranged from crushed firetrucks to elevator number placards. These items would become the backbone of the 9/11 Museum collection and were moved again to Hangar 17 at John F. Kennedy International Airport under the purview of the Port Authority of New York and New Jersey.

It was not until 2006, that these recovered artifacts began coalescing into a cataloged museum collection, along with numerous other items including:

- personal objects and ephemera donated by victim's family members, survivors, first responders, witnesses, and other stakeholder groups
- materials from spontaneous memorials
- tribute and response art
- archives, archeological remnants, and memorabilia related to the significance of the original World Trade Center site
- audio and video files
- print and digital images
- oral histories

Documentation and cataloging began in earnest with the onset of plans to develop recovered material into a museum. Material characterization at this stage was limited to defining the primary components in broad categories: textile, paper, metal, ceramic, plastic, etc. By 2023, 4,000 objects were labeled in the collection management database as "plastic." To find plastic objects potentially not classified in the original inventory, object categories such as boots, sunglasses, personal protective equipment, and identification cards were searched within the collection database since these were most likely to contain plastics. Almost 5,000 additional plastic-containing objects were identified for a total of 8,820 (or 38% of the entire collection). This new estimate includes objects that may have plastics as minor components. For example, the quilt collection in which many personalized quilts contain tribute items like beads, laminated photographs, and artificial flowers.

Complicating the review of the plastic collection is that recovered artifacts from the three crash sites (the World Trade Center, Pentagon and Shanksville, Pennsylvania) exhibit damage as a testament to the trauma they endured. In the case of the plastic materials, this damage can be difficult to distinguish from material degradation due to aging, and in some instances the extreme conditions of their 9/11 provenance (i.e., heat, water, pressure) may be accelerating their deterioration (Figure 8.3).

Further, objects from Ground Zero may be covered with "World Trade Center dust" and debris obscuring surface details and complicating material identification. Due to the known toxicity of the dust, once they enter the collection, recovered artifacts from all sites are considered hazardous until cleared through analytical testing. Handling, storage, and display of these objects are dictated by strict collection management guidelines that prioritize the health and safety of anyone who may encounter them. This

(a) (b)

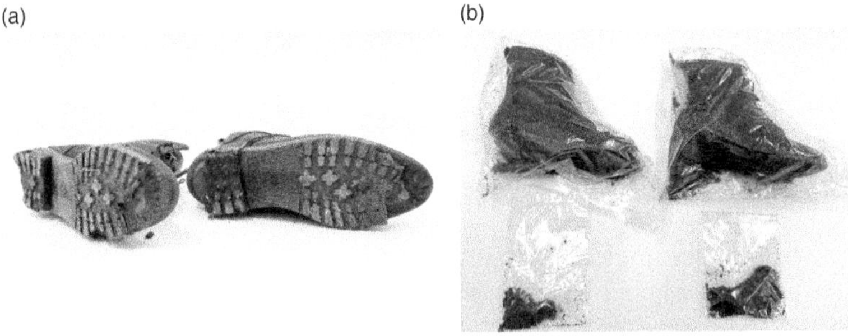

Figure 8.3 Boots (C.2014.480.1) worn by Carol Orazem on September 11, 2001. (a) as they appeared upon acquisition in 2014 and (b) in 2024 in plastic enclosures, note severe deterioration of the soles. Credit: Collection 9/11 Memorial & Museum, Gift of Carol Orazem-DiPaolo.

includes storage in sealed plastic enclosures. Artifacts may only be viewed or handled outside of these containers by trained personnel with proper isolation measures, engineering controls, and personal protective equipment (PPE). In addition to potentially hastening deterioration, these containers limit access for accurate examination, documentation, and material identification (Figure 8.3).

The survey revealed that within the collection of 3,500 "recovered" artifacts, 2,160 plastic artifacts are isolated and awaiting testing for hazardous materials. Previous testing data demonstrated that 54% of plastic-containing artifacts suspected to be hazardous by virtue of being "recovered" are no longer deemed hazardous following testing. Once cleared, these objects can be removed from their enclosures and examined more closely. Therefore, it was determined that analytical testing to clear as many objects from the hazardous materials protocol would be a necessary part of any future plastics condition survey.

The pilot survey was also instrumental in identifying types of plastics to target for materials testing, condition assessment, and rehousing. Prioritized plastics included types exhibiting signs of structural instability or those with indicators of imminent continued degradation (e.g., weeping, discoloration, smell). Given the age of the collection and the categories of objects, PVC and elastomers were suspected to be the primary problematic plastics, which was confirmed during the survey. The most pressing issues noted were sticky surfaces, embrittlement, cracking, crumbling, and deformation of elastomers.

Using a customized survey form developed specifically for the needs of this collection, the condition surveying of 112 objects (1.3% of the plastics collection) was completed in 45 hours. It confirmed that over 50% of the plastics are actively degrading and in poor or fair condition; supporting that implementation of the collection-wide survey is urgent. Since differentiating between the deterioration of plastics and the damage that occurred from the events of 9/11 will only become more difficult, virtual preservation through 3D imaging was also identified as a priority for the plastics collection.

Additionally, 64% of objects surveyed require rehousing. Special considerations, such as isolation, ventilation, enclosures, cool/cold storage, dark storage, anoxic storage, or use of scavengers/adsorbers for storage were recommended for 38% of objects surveyed; none of which the museum currently has the resources or capacity to implement.

From the results of this initiative, staff determined that the entire plastics collection could be surveyed over a two-year period with the addition of a conservation position hired specifically to manage and conduct the survey with the assistance of the current full-time staff of three people. In addition, the purchase of a FTIR spectrometer would help accelerate material identification along with a workshop to help all museum collection staff classify plastics going forward since the collection continues to grow. A comprehensive survey, estimated to cost $350,000, would include material identification, photography, and condition assessment for all 8,820 objects plus hazardous materials testing, photogrammetry, and rehousing for a selection of the most at-risk artifacts.

Labeling

Poor decisions of how to label plastics can hasten deterioration since solvents, plasticizers, and pigments or dyes in the marking pen, barrier coat, or label adhesive can negatively interact with the plastic (Fenn 1993). Overtime this may result in damage such as blistering (Figure 8.4), discoloration, crazing, change of gloss, or the ink wicking into the plastic making the label illegible (Lockshin 2019).

When labeling, it is not always possible to avoid solvents. So, is there a way to choose a solvent that is less likely to harm a plastic? Each type of plastic has a value known as a solubility parameter which can be used to determine if a plastic is soluble in a solvent as noted by the plastic swelling, softening, or dissolving (van Oosten 2022). Generally, since like dissolves like, a polymer will dissolve in a solvent that has a similar or lower solubility parameter or if the polymer and solvent are within 2.0 units of each other (Shashoua 2008; van Oosten 2022). For example, acetone, which is a solvent sometimes used in lacquers for labeling, as seen in Figure 8.4, has a Hildebrand solubility parameter of around 20.0 meaning that CN, polymethyl methacrylate (PMMA), PVC, polystyrene (PS), polyethylene terephthalate (PET), and polycarbonate (PC) are soluble in it (Shashoua 2008; van Oosten 2022). Since it may be difficult to identify the type of plastic and there are so many that are negatively affected by acetone, it should NEVER be used for labeling plastics. It may be tempting to consider ethanol as a less aggressive

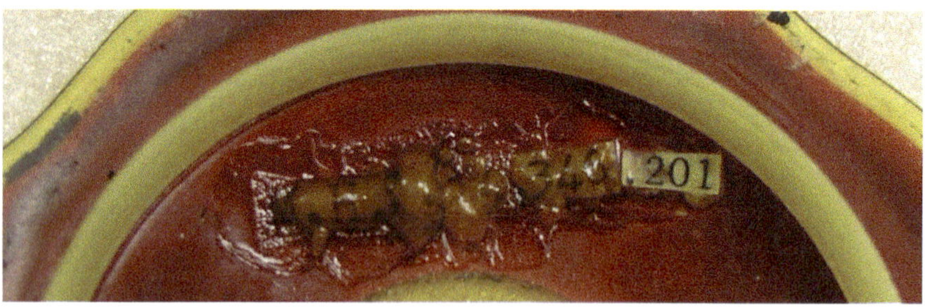

Figure 8.4 Catalog number applied to CN by printing on paper and coating with an acetone lacquer. Solvent interaction caused blistering, making the label illegible. Deaccessioned. Photographed by author, Division of Medicine and Science, National Museum of American History, Smithsonian Institution.

solvent because its solubility parameter is pretty high at 25.4–26.0, meaning that most plastics are more than 2.0 units removed from it (Shashoua 2008; van Oosten 2022). However, labeling studies from collection observation and testing have found that alcohols can harm PC, PMMA, CN, CA, and PS (Fenn 1993). This is because a plastic's reaction with a solvent can be affected by other factors like its degree of degradation and the formulation of that specific plastic can cause the solubility parameters to differ from what is expected for that plastic family (van Oosten 2022). The solubility parameters attributed to solvents themselves can also vary slightly between published sources making it less than precise to compare parameters for solvents and polymers. So, it is best to treat solubility parameters as a general guide (Shashoua 2008).

Neutral water-based adhesives, lacquers, and inks may seem like the answer to the solvent threat but remember that water is a solvent. The Hildebrand solubility parameter for water is 46.8–47.7 (Shashoua 2008; van Oosten 2022). Certain plastics like CA and CN are sensitive to water, as discussed in Chapter 4, so its use can cause damage. For some plastics though, water-based labeling solutions may be okay (Fenn 1993).

To avoid potential solvent interactions all together, indirect labeling is preferable when possible. This generally means tying on tags made of either paper or Teflon tape, sometimes called plumber's tape. Though safer chemically, detachable tags come with the risk of dissociation since if the tag is lost or damaged, the object identification is lost. As some plastics deteriorate, their degradation products may be adsorbed by the tags or ties making them acidic, brittle, and prone to failure and loss (Fenn 1993).

If direct labeling is required, try to label a non-plastic component. Or if that is not an option, a soft graphite pencil (2B–4B) may be used but take care not to dig into the surface. It is also important to choose an inconspicuous area for labeling that is protected from light and not likely to be stressed from manufacture, meaning that an injection point, curved edge, or near a maker's mark are areas to avoid (Fenn 1993).

Sidebar 8.7 Labeling Textiles

Kris Cnossen

"Stitched Label Method" (based on method used at Winterthur Museum, Garden & Library)

Materials:

- Pigma Micron Pen in size 1 in black ink
- Iron
- Clean water, preferably deionized or distilled

- Plain woven or twill cotton tape – 5/16″ wide
- Thin thread of an appropriate color or undyed (hair silk is a good option)
- Needle (size 12 beading is best)

Method:

1. Cut tape long enough for accession number plus 3/8″ on either end to tuck under.
2. Write accession number on tape. Pinning the tape to a piece of board can help in keeping the tape steady while writing.
3. Set the ink by placing the tape between two layers of clean muslin, setting the iron to "cotton," and ironing the tape and muslin for 30 seconds on each side.
4. Wash the tape by hand using water and very small amount of Orvus detergent. Rinse thoroughly.
5. Allow tape to air-dry or use a hot-iron to dry the tape faster.
6. Sew the label on the textile in the designated area. Whip stitch the edges, being careful not to use too many stitches and being aware of what will be seen from the face.
   a) The label should be hidden when viewed from the front, but close enough to an edge to be easily found.
   b) It is common to add a museum label to textiles in an area where a label would be found (i.e., the back of the neck of a shirt).
   c) It is helpful to create a standard label placement as an institution for textiles, such as the back, proper right side.
   d) Try to sew into seam allowances and take care to not cover any pre-existing labels.
   e) Do not sew into leather, suede, or elastic as this causes unnecessary damage. In these cases, use a tie-on label or similar labeling system used for other plastics.

Key Points for Plastics Collection Management Part 1: Processes and Procedures

The processes and procedures for handling and labeling plastics have consequences for how they age and are vital points of overlap between preventive conservation and collection management. The decisions made regarding how surveys are conducted and information is documented in the collection database can help ensure that insights about plastics can be accessed for future use, such as determining the degree of degradation over time and the probable cost of care.

References

AICCM (Australian Institute for the Conservation of Cultural Materials). n.d. "Visual Glossary." Australian Institute for the Conservation of Cultural Materials. Accessed May 1, 2024. https://aiccm.org.au/visual-glossary/

Bacci, Mauro, Costanza Cucci, Thea van Oosten, Yvonne Shashoua, and Jozef Rychly. 2012. "Surveying Plastic Artefact Collections in Museums." In *Preservation of Plastic Artefacts in Museum Collections*, edited by Betrand Lavédrine, Alban Fournier, and Graham Martin, 139–149. Paris: Comité des Travaux Historiques et Scientifiques.

Barker, Claire S. 2010. "How to Select Gloves: An Overview for Collections Staff." Conserve O Gram 1/12. National Parks Service. www.nps.gov/museum/publications/conserveogram/01-12.pdf

Bell, Julianne, Karen M. Thompson, Karina Palmer, Gavan McCarthy, Melanie Barrett, Ellen Burrows, and Petronella Nel. 2022. "Collaborative Development of Polymer-Based Collection Survey Methodology and Relational Data Model." *Studies in Conservation* 67 (sup1): 10–22. https://doi.org/10.1080/00393630.2022.2049034

Cato, Paisley S. and R. Robert Waller. 2023. "Agent of Deterioration: Dissociation." Canadian Conservation Institute. www.canada.ca/en/conservation-institute/services/agents-deterioration/dissociation.html

Design Museum Gent. 2022. "Thesaurus for Plastics: Materials." www.designmuseumgent.be//assets/content/plastics/Thesaurus-for-plastics-materials.pdf

Fenn, Julia. 1993. "Labeling Plastic Artefacts." In *Saving the Twentieth Century: The Conservation of Modern Materials*, edited by David W. Grattan, 341–350. Ottawa: Canadian Conservation Institute.

Hatchfield, Pamela B. 2002. *Pollutants in the Museum Environment: Practical Strategies for Problem Solving in Design, Exhibition, and Storage*. London: Archetype Publications.

IPI (Image Permanence Institute). 2024. "FilmCare.org, Visual Decay Guide." Image Permanence Institute, Rochester Institute of Technology. www.filmcare.org/visual_decay

Keneghan, Brenda. 2001. "A Survey of Synthetic Plastic and Rubber Objects in the Collections of the Victoria and Albert Museum." *Museum Management and Curatorship* 19 (3): 321–331. https://doi.org/10.1016/S0260-4779(01)00054-1

Krieg, Till, Cristian Mazzon, and Elena Gómez-Sánchez. 2021. "Material Analysis and a Visual Guide of Degradation Phenomena in Historical Synthetic Polymers as Tools to Follow Ageing Processes in Industrial Heritage Collections." *Polymers* 14 (1): 121. https://doi.org/10.3390/polym14010121

Ledoux, Nicole, Georgina Rayner, Susan Costello and Angela Chang. 2023. "Preventative Conservation, Treatment and Technical Study of Plasticized Poly(vinyl chloride) Multiples by Joseph Beuys." *Studies in Conservation* 68 (3): 343–356. https://doi.org/10.1080/00393630.2022.2033520

Lockshin, Nora Sharon. 2019. "Marking Collections." In *Preventive Conservation: Collection Storage*, edited by Lisa Elkin and Christopher A. Norris, 781–806. New York, NY: Society for the Preservation of Natural History Collections.

POPART (Preservation of Plastic ARTefacts in museum collections). n.d. "Damage Atlas: Atlas of Case Studies Presenting Typical Damages." POPART. https://popart-highlights.mnhn.fr/wp-content/uploads/3_Collection_survey/5_Damage_atlas/Damage_atlas.pdf

Shashoua, Yvonne. 2008. *Conservation of Plastics: Materials Science, Degradation and Preservation.* Oxford: Elsevier/Butterworth-Heinemann.

Stephens, Catherine H., Eric M. Breitung, Corina Geiculescu, and Thomas Strange. 2022. "Determination of Nitrile Gloves Appropriate for Use When Dry Handling Art." *Journal of the American Institute for Conservation* 61 (3): 162–171. https://doi.org/10.1080/01971360.2021.1913543

Taylor, Joel. 2014. "The Impact of Assessment Guides on the Reliability of Collection Condition Surveys." In *Preprints ICOM Committee for Conservation 17th Triennial Meeting Melbourne Australia 19-23 September 2014*.

van Oosten, Thea. 2022. *Properties of Plastics: A Guide for Conservators*. Los Angeles: Getty Conservation Institute.

Wessel, Thomas. 2008. "You Can Have a New One – It'll Be Better Anyway! Plastic Challenging Art Insurance." In *Plastics Looking at the Future and Learning from the Past – Papers from the Conference Held at the Victoria and Albert Museum, London, 23–25 May 2007*, edited by Brenda Keneghan and Louise Egan, 138–141. London: Archetype Publications.

9 Plastics Collection Management Part 2

Discussions and Decisions

This second part of the collection management-focused content addresses the considerations that go into decisions around accessioning plastics into the collection, putting them out on exhibition, including for loans, and when replacement, replication, or deaccessioning may be prudent. There are many nuanced factors that go into these discussions and the decisions that are reached may differ between institutions, even if the plastics are similar, because of the museum's mission, curatorial norms and practices, and the degree to which the anticipated lifespan and associated potential risks of a specific class of plastic are understood and factored.

Accessioning

As evidenced in preceding chapters, especially Chapter 4, many plastics will age in ways that mean they will not look the way they did when accessioned, even with sound care in stable, or relatively stable, environmental conditions. This is often because plastics are made with anticipated lifetimes for use. For example, back in the mid-1950s, industry defined the point at which plasticized polyvinyl chloride (PVC) film was no longer useable as when it lost 10% of its original weight as plasticizers and other additives migrated out, but the time period for that degree of change was expected to vary from months to 1,000 years depending on formulation and environment, particularly temperature (Shashoua 2008). Or, in other words, manufacturers knew their plastics would not last, but their actual lifetime depended on additives and environmental conditions. The earliest PVCs are approaching the century mark from their creation (Figure 3.6). How they have aged can vary greatly, with collection surveys reporting that degradation can become evident anywhere from 5 to 35 years after entering the collection (Shashoua 2008). Unfortunately, before accessioning, it is not usually possible to predict which plasticized PVC will age poorly because of all the variables of manufacture, environmental exposure, and stresses that occur before it grabs a curator's attention.

It may be tempting to exclude the problem plastics from being accessioned in the first place. This would alleviate concerns of degrading plastics negatively affecting other materials and save time and energy of collection staff since plastics of concern should be periodically monitored, as discussed in Chapter 7, and may

DOI: 10.4324/9781003489658-9

require specialized care in terms of storage materials, increased ventilation, or kept at low temperature, as covered in Chapters 5 and 6. For some, the care needed for plastics may outweigh the value they bring to the collection. And in the case of low-temperature storage, which requires funds to start up and maintain as well as uses a lot of energy to run, some may argue that this type of care is counter to sustainability efforts that cultural institutions should be supporting. But if these plastics are not accessioned, what gaps in material culture could result, and how could that affect future research and understanding? Whose artistic practices would be excluded (Sidebar 9.2)? What if future discoveries for interventive conservation treatment or preventive care extend the lifespan of these plastics, but they have since been lost because they were not accessioned?

Perhaps a more reasonable approach is to adjust expectations. In collection care, we strive to prevent change once an item is accessioned, as if it is possible to halt the passage of time. Markers of good care for stable materials are often set at a minimum of 100 years without marked change if kept in a stable environment (Elkin and Waller 2019). But is this actually what occurs? Are yellowed papers considered damaged, or is this an acceptable sign of aging? Why is some metal corrosion accepted as a patina of age? If an oil painting has an even craquelure and the paint is not actively flaking, can it be exhibited, or would visitors be distracted by the fine network of cracks and assume inadequate care? Once you stop and think about many materials in collections, changes from their original state are often readily accepted by museum staff and visitors and sometimes add prestige (Shashoua 2008). So, why are plastics different? Is it because they are relatively new? Are we so used to them in our daily lives and treat them as a disposable commodity that we never really see them age? Do we think of them as replaceable and not attribute as much value? Perhaps it is all of these factors that influence how we view plastics and expect to view them once in a museum. But what if plastics are accessioned with an anticipated life expectancy? Could that guide collection preservation and access decisions?

The Science Museum, London embarked on an effort to identify problem plastics (cellulose nitrate (CN), cellulose acetate (CA), PVC, and polyurethane (PUR) foam) in their "acquisition scorecard," which seeks to estimate conservation and care costs anticipated over the next decade before an object is accessioned (Coles 2008). Since familiarity with plastics can vary and anticipating needs can be difficult, estimating the cost of future care is trickier with plastics than for traditional materials that have decades or even centuries of amassed preservation knowledge to draw upon. The approach the Science Museum took was to develop fact sheets that for each problem plastic included:

- Period of production
- Uses
- Deterioration
- Images of typical degradation

These fact sheets also included flow charts to confirm how identification was made, which proved most needed when the object was in good, stable condition since it is often easier to work backwards using degradation signs to identify the plastic (Table 4.2). Identification drew upon dating the object, which can be supplemented with timelines like Figure 3.6 or those readily available online, as well as use and characteristics such as flexibility and odor. If identified as a problem plastic, tailored storage recommendations, including the need for ventilation or lower temperature or relative humidity (RH), were noted. If a problem plastic had visible deterioration, the recommendation was to not accession into the collection (Coles 2008). In reporting on this project, the conservator noted that curators were conflicted by the desire to collect, since that is often a key part of their job, and concerns about "...leaving a negative legacy of deteriorating objects..." (2008). Discussions with the curators brought up concerns of whether fears that something may degrade would exclude the accessioning of significant objects or if not acquiring objects that can pose risks is good stewardship. In the end, most of the curators concluded that even with a limited lifespan, if significant enough, a problem plastic is still worth collecting, but they should be used "... to their full potential..." in the time that they are exhibitable (Coles 2008). Whether that holds true for your collection and whether that view changes over time can only be known if the discussion is started, which hopefully readers will feel empowered to do.

So, how much degradation is acceptable in a plastic? Does it vary by the type of plastic or the form it takes, say, an artwork versus a commodity product accessioned to show daily life in the 20th or 21st century? At what point is the change so severe that the plastic is no longer recognizable or has diverged so far from its original form or intent that it can no longer tell a story or provide insight to researchers or the public? Can the degree of degradation be projected using environmental data for the storage or exhibition spaces they will likely be kept in for long periods of time? And can these data be used to anticipate potential issues before accessioning? And, if accessioned, guide decisions for exhibitions and loans?

The Image Permanence Institute developed the Time Weighted Preservation Index (TWPI) as a measurement for estimating the effects of temperature and RH over time on carbon-based organic materials made from plants, animals, paper, and plastics until they show "significant deterioration" as marked by "noticeable discoloration or embrittlement or other changes that involve a serious loss of appearance or functionality" (Nishimura 2011). Since its introduction over 20 years ago, the TWPI has become an industry standard and is a utilized metric in many environmental monitoring products in order to gauge how organic materials will age in the environmental conditions being tracked (Shpargel 2003; Conserv 2024). The higher the TWPI, the slower the anticipated rate of deterioration. Since high temperatures accelerate chemical reactions, the calculated differences between no air conditioning and air conditioning are about 20 years, while cool storage can earn a TWPI of 70 years, and cold storage can yield more than 250 years (Shpargel 2003). If at least a year's worth of data is available for a space, the TWPI could be used to help gauge likely aging of materials, including plastics (Conserv 2024). Obviously, unknown variables in the plastic's formulation and past environmental conditions affect aging. But TWPI may be an available indicator at some institutions that could factor into decisions about whether or not to accession a plastic. For example, if the storage room that a newly accessioned CA object would be placed in has a consistently low RH and the TWPI is several decades, there may be better confidence in acquiring it than if it was entering a space with wide fluctuations or a high RH and high seasonal temperatures. Or is there value in the acquisition regardless of the potential for storage conditions to negatively influence how long the plastic is expected to remain exhibitable? The answer likely lies in how the object fits into the museum's mission and the people making the decision. Or, in other words, even if efforts are made to make data-driven decisions, there is an inherent element of subjectivity in museum acquisitions.

Discussions about whether the expected useable lifespan of a plastic should be a factor when accessioning, the potential impact that could have on the types of works collected by the institution, and the possible exclusionary practices that may result are conversations that should occur among curatorial and collection care staff and other stakeholders. The rationale for the decisions reached should be recorded and periodically re-evaluated.

Exhibition and Loans

Once the decision is made to accession a plastic, the discussion needs to pivot to determining parameters for access. If the finite lifespan of plastics could be a factor in accessioning, should it also be part of decisions for exhibitions and loans? If a plastic is fragile, is the goal to try to halt change as much as possible, perhaps in low-temperature storage, and have the plastic available to researchers and curators with advanced planning while very occasionally permitting short-term exhibition in well-controlled exhibit environments? In this approach, preservation trumps access, and some may think that should be the goal of all collection care. Or is there equal validity in efforts to increase access? Does it better fulfill the mission of the institution to exhibit the plastic, whether in the museum or loaned to another, so more people can see it before it has potentially altered so far from the maker's intent or original use that little can be gleaned from it? Since without interventive conservation treatment, a plastic will never look better than it does at this moment, do museums have an obligation to make them more accessible, whether that is to researchers behind-the-scenes, put on exhibition at their home institution, or approved for requested loans (Sidebar 9.1)?

Or does the answer depend on the plastic object and the specific request for exhibition? For example, if a problem plastic is acquired and immediately requested for loan and there are concerns for damage during transport and from light exposure while on view, should the loan be refused and the plastic kept in storage? Does the answer change if the loan request is to include the plastic in an artist's retrospective? If the artist worked in plastics, is the window of opportunity closing to bring the works together while they still resemble their original forms? What is known about how the artist viewed changes to the material over time, and should that inform decisions? What if the artist's opinions change over time and they felt one way at creation and another way decades later? Or, as in the case with Eva Hesse, who knew that her latex works would not last and said, "Life doesn't last; art doesn't last," free decision makers from the debate of whether it is okay to display a drastically changed work (Guggenheim 2022)?

Sidebar 9.1 Loan Request of Deteriorating Plastic

As a conservation fellow focused on plastics in 2005 at the Smithsonian's National Museum of American History (NMAH), I was asked to review a loan request to borrow Spock's ear tips from the 1982 movie *Star Trek II: The Wrath of Kahn* (Figure 9.1). The pointed tips were likely made from PUR foam and would have been adhered to the actor's ears while filming (Barone 2018). The NMAH ear tips were accessioned in 1999 from a donor who acquired them in 1987 (NMAH n.d.). Over the 23 years from when the movie came out to when the loan request came in, the ears had darkened unevenly and small fragments readily crumbled from the edges, as is typical of aged PUR foam. Because of the discoloration and fragile condition, I advised that the loan be denied and the ears be returned to storage. Though that was a common mindset then, it is not a decision I would reach today. As PUR foam that was entering its second decade, the ear tips were coming to the end of their anticipated useable lifetime. Having them rest in standard storage did not halt deterioration. Now, 20 years after the loan request and over 40 years after filming, the ears are even more fragile, and the museum has recently 3D scanned them to facilitate the creation of a more supportive storage mount (Ryan Lintelman, Curator, NMAH. Personal communication. April 29, 2024). Given the increased fragility, the insight and joy the ear tips could have brought to people at the exhibit decades ago now seem like a missed opportunity.

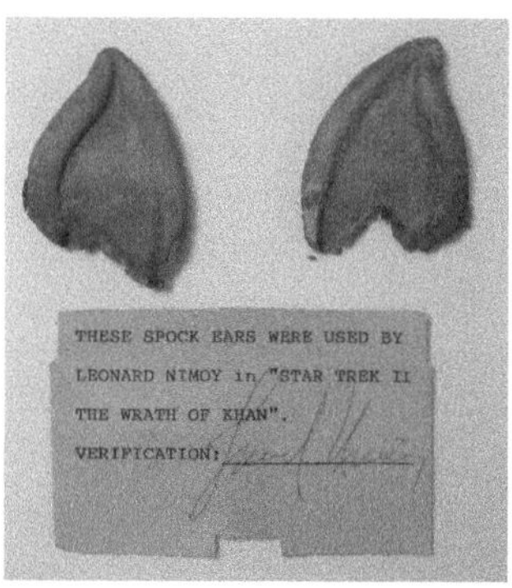

Figure 9.1 Spock ear tips from *Star Trek II: The Wrath of Kahn* photographed in 2005 as part of a denied loan request. NMAH 1999.0102.01/.02. Photographed by author, Division of Culture and the Arts, National Museum of American History, Smithsonian Institution.

Replacement or Replication

Those working with plastic collections may have to make difficult decisions about whether the primary goal is to preserve the actual object or preserve its form through replacement with a similar object or replica. Views about this and the conclusions reached will undoubtedly depend on the type of collection and the specific object being evaluated. If the plastic is a consumer product, such as an early piece of Tupperware that is not aging well, but one of the same design that appears to be in better condition becomes available, should it be purchased as a replacement or back-up for the accessioned one? Does the answer depend on the story the object was accessioned to tell? If it was collected with the intent to convey life in 1950s American suburbs or as a general example of increased dependency on plastics over time, would it be better to replace or augment the deteriorating one? If the object was collected because of its association with a specific person or historical connection, does replacement seem too radical? Or in other words, how much does the aura of the original come into the decision processes? And at what point do the limitations of the material impact those decisions?

If the plastic is part of a unique artwork, how does that influence the conversation around replication? When does its transformation as it degrades erode the artist's intent to the point that showing it may be undesirable? How much does that change if the maker understood and embraced this quality of aging, such as with Eva Hesse discussed above? What if the artist made multiple versions in anticipation of the plastic aging, as discussed in Sidebar 9.2? When is it acceptable to produce a replica of the work in order to convey the intended form? Should the same problem plastic be used knowing it has a finite lifespan, as the artist Naum Gabo did when an earlier work of CA degraded, so he made a replica of it again in CA 30 years later (Lodder 2007)? Or is a more stable plastic of similar appearance appropriate to use? Does working in a different medium still capture the artist's vision, and can that be exhibited, or is it rather a way to document the original? This approach was taken by the Philadelphia Museum of Art with Gabo's CA work *Construction in Space: Two Cones* when they "...commissioned a copy in Plexiglas from a model-maker to keep as a record of the decayed original" (Lodder 2007). If the decision is made to use a different plastic, should it follow the notion that "In changing a material it should be that science and not eyes could detect the difference" (Williams and Williams 2007)? What happens to the original degrading plastic artwork? What about the reproduction and any subsequent ones? How are they marked as replicas? Do they actually replicate the original, or are replicas, especially by another hand, inherently reinterpretations? If one or more replicas are made over time, are they all accessioned and stored together? If displayed, how much should the public know about these decisions? Which is more authentic – the artist's original work that is degrading or the replica meant to capture the art's original form? Is the visitor experience impacted by reproductions, or could the display of degrading materials alongside replicas serve as an added layer of discussion?

Sidebar 9.2 An Artist Interview Case Study Highlighting the Significance of Plastics in Shervone Neckles' *Domiciliation: Bless This House*

Steven O'Banion and Samantha Owens

In the following interview, Steven O'Banion (SO) and Samantha Owens, both conservators, sat down with Shervone Neckles (SN) in her studio to talk about the use of plastics in her oeuvre, using the work *Domiciliation: Bless This House* as the primary example. The artist and interviewer had a pre-existing relationship, having known one another for several years. Samantha Owens video recorded the interview in front of elements from the artwork and several archival papers and images. The intended use of the interview, to be included in this publication, as well as a list of topics that they hoped to cover, were shared in advance.

The discussion started with the artist introducing her artistic practice, history, and influences. She noted,

> A lot of the mediums that I use, like textiles and assemblage and printmaking and sculpture and multimedia installations and now even public art, they're weaving together my Afro-Caribbean, Grenadian aesthetics and expression and sensibilities with my American identity, and then also my interest in history, nature, and the sciences. I think, at its core, really what my work is about is honoring and celebrating my ancestors and making visible their work and their labor and ingenuity that might otherwise be lost or forgotten.

The conversation then proceeded to discuss her *Domiciliation* series, commissioned for the Grenada Pavilion of the 2019 Venice Biennale, consisting of three-dimensional works fabricated in 20-gauge PVC with screen printed images. She explained,

> My grandmother was a seamstress, but she would also do a lot of upholstery work, so she would use that material very frequently. As I'm using this, I'm referencing not only how we preserve things, but also the material she used in her seamstress work. It has layered meanings in this piece for me.

The interview then moved onto the preservation of this body of work. The following transcript has been edited and condensed.

SO: I might pivot a little bit and start talking about the preservation and the extended life period of the artwork. As a starting point, have you noticed changes in the artwork since it was originally fabricated?

SN: Mostly in this one [*Domiciliation: Repository #1*]. This piece here [pointing to the artwork]. It's only because of how it's hung. I've learned a couple things about hanging these works and where to find the tension points. So, I'm going to go back in and actually open up the sides and resew the area and release where the tension is currently stretching the material. That's the interesting thing about when you make something and you have a very particular vision for how you want it to be presented.

I forgot to explain the steps of the fabrication process. After silk-screening the collages onto the PVC surface, I then took the PVC prints to a custom sewing company and I had them sew all of the prints together with gold trim and boning. Following that, I reinforced the pieces by adding metal rods to ensure the artwork would hold up and it wouldn't just collapse onto itself. Each of those steps occurred separately. Ideally, when planning, I would prefer to have everyone (my collaborators) in the same room discussing the project altogether and making suggestions like, "Hey, you might also want to think about how the plastic stretches when you hang the work this way." The plastic in *Domiciliation: Repository #1* did stretch a bit, so I had to tack the area a bit more where there was a slight warp happening. It was a structural detail that I needed to address immediately.

In terms of yellowing, I don't anticipate that happening with this body of work, even though yellowing is what normally occurs over time with plastic. But I'm definitely controlling that process by protecting the work in custom-made archival boxes and storing them in an environmentally-controlled storage facility. By doing this, I know I'm slowing that process down tremendously. I also intentionally used a high gauge and high-quality PVC material to ensure the long-term preservation of the work. I don't foresee yellowing or any signs of deterioration happening within my lifetime. But even if it does begin to show signs of aging, it's okay because, much like the subjects represented in this body of work, all things age. Just as the paper in my grandmother's journal ages, my ancestral home continues to age, these pieces will also experience natural changes over time.

For me, the goal is for the work to be acquired and presented thoughtfully, with care and preservation in mind. The pieces need to be given breaks. They can be exhibited over a period of months in a dimly lit environment. The lights should be turned off when the museum/gallery is not in operation, so the work can rest. After a certain amount of time, they are stored away in complete darkness until the next presentation.

SO: And conceptually that's a really key fact. The plastic isn't representing newness, it's representing durability.

SN: Yes, that's right. That's precisely the point I'm trying to make here. Especially given the frequency of hurricanes impacting the Caribbean, preparedness and preservation are always top of mind. That's why I made a second edition, just in case anything were to ever happen to the first edition. The second edition is slightly different from the first; it features an upholstery trim similar to the kind my grandmother would use to upholster furniture. So essentially, the second edition is designed to serve as a backup.

SO: When you put so much effort into a work that represented you at such an important part of your career, you must be thinking about where they get placed in the long term, whether that be with an institution, an individual. Did that play into your considerations and making two separate versions?

SN: Absolutely. It's one of the things where, because a lot of my work has so much connection to family, that it comes up a lot for me about where the work will live beyond me, and what would serve my family's continued story. Where would my family, my grandmother, where would they be proud for our story to be told? Where would others, who can relate to our story, see themselves? Where do they feel connected to? Where and who preserves our stories in our own voices? These are the questions I'm constantly thinking about. I know these are the same questions that will help me decide where I want the work to be placed when that time comes.

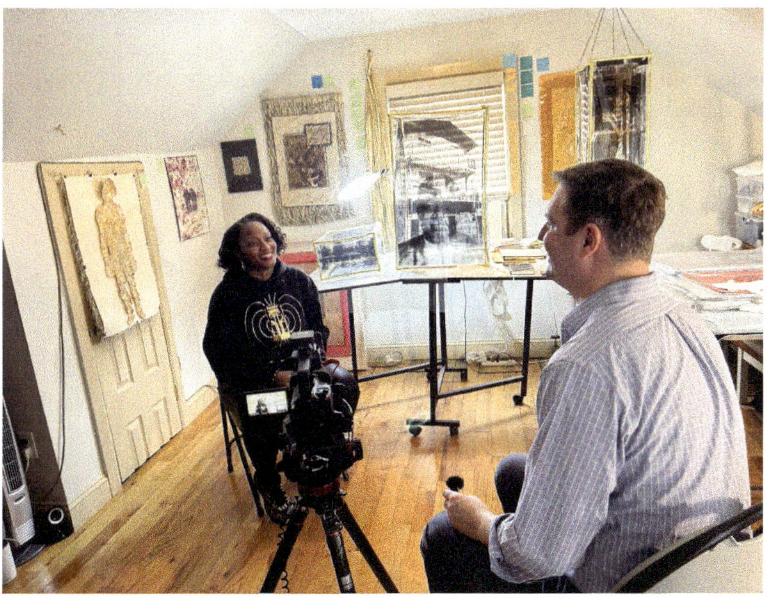

Figure 9.2 Steven O'Banion and Shervone Neckles in the artist's studio during the interview in front of elements from *Domiciliation: Bless This House* and other works. Photographed by Samantha Owens.

SO: As we're wrapping up, is there anything you wanted to say that I missed?

SN: Ohh, the title of the work, *Domiciliation: Bless This House*, is important to explain. So, the term "domiciliation" means making a country your legal home. The subtitle, *Bless This House*, is actually my grandmother's and my great-grandmother's favorite gospel song, particularly the 1957 rendition by Mahalia Jackson. It has become a family heirloom that connects the generations of matriarchs in the family. Much like the lyrics in the song, the work expresses the sacredness of home, asking for blessings for those within the home, entering the home, and all those carrying home within them.

Deaccessioning

Until the last decade or so, deaccessioning was a dirty word in many collecting institutions, and few wanted to support pruning collections because who are we to decide if something will be informative or of value in the future? If it is in the collection, isn't there an obligation to keep it? For most museums, collecting has out-paced available storage and attempts to keep that storage in controlled temperature and RH ranges, even when broadened from the previous tight set points, have put a strain on aging heating, ventilation, and air conditioning (HVAC) systems, increased costs, and has been scrutinized by growing efforts to make sustainability essential to ethical collection care. Add that many institutions have plastics in their collections that are severely degraded; it has to be asked: why are we holding on to some plastics and when do we let them go?

In the framework of this book, deaccessioning is in the context of disposal and destruction because of degradation and loss of value as opposed to an accessioned object that is out of scope of the museum, redundant, or lacks value to the institution and is transferred to another institution or sold. In deaccessioning guidelines, whether from a national body like the American Alliance of Museums (AAM) or an international one like the International Council of Museums (ICOM), provisions for deaccessioning because of poor condition are specified. For ICOM, the first three deaccessioning factors are tied to condition and specialized needs, which can obviously apply to plastics (2019):

1. Objects that are damaged beyond reasonable repair and are of no use for study or teaching purposes may be destroyed.
2. The object poses threats to health and safety to the staff and the public.
3. The museum is unable to care adequately for the object because of its particular requirements for storage or conservation.

If a decision is reached to deaccession a plastic, the next steps are formal disposal from the collection and destruction. In "ICOM Code of Ethics for Museums," the section on disposal from collections notes that

> The process of deaccessioning and disposing of an object should be carefully documented in the museum's collections records, and these records should become a permanent part of the museum's records, maintained even after an object is disposed from a museum's collection.
>
> (2017)

This ensures that the record of the object's presence in the collection remains even after it is physically removed. How the plastic is destroyed may vary at institutions but typically involves physical destruction and throwing away the fragments. Because some plastics may be a hazard as they break down, how the plastic waste is actually disposed of may depend on environmental rules of a specific jurisdiction (Sidebar 2.4). For example, CN film should not be discarded with regular trash because of its fire risk and PVC should be considered hazardous waste because of its potentially harmful additives, as discussed in Chapter 4.

It is worth noting that the U.S. National Park Service specifies that CN films in their collections do not require formal deaccessioning when reformatted. Rather, the new, stable, reformatted image replaces the unstable CN film as the museum object and the CN film can be destroyed and disposed of, assuming there is not a unique value inherent to the original physical film (Floray 2004).

In the example outlined above about accessioning plastics at the Science Museum, London, the curators who supported accessioning known problem plastics did not support deaccessioning because degradation of the material now told a different story (Coles 2008). As a science museum, one could see how the chemistry of plastic degradation could fit into the museum's mission. But does that logic hold at other types of museums? Again, the only way to get answers and establish protocols is to start discussions with a range of stakeholders and to periodically re-evaluate the decisions that are reached.

The process for deaccessioning will vary by each institution and should be outlined in the museum's Collections Management Policy. The decision to permanently remove an accessioned object from a collection should never be the decision of one person. That said, it is often a collection steward who is best situated to flag plastic issues as objects are pulled for loans or for researchers, or as inventories or condition surveys are undertaken. Understanding the consequences of keeping a degrading plastic in the collection in terms of potential impact on other materials and the plastic's requirements for care are essential factors to bring forth in deaccessioning discussions. The holders of that knowledge tend to be those specializing in collection care and, therefore, may be the ones who are asked to comment on the validity of deaccessioning criteria based on materiality. With the growing number of plastics in collections, it is increasingly important that collection stewards understand plastics more fully.

Sidebar 9.3 Deaccessioning Deteriorating Plastics

Hundreds of mathematical models dating from around 1915–1950, composed of paper and plastic, made by A. Harry Wheeler, a high school mathematics teacher from Worcester, Massachusetts, and his students, were accessioned by the Smithsonian's NMAH in 1979. By the turn of the 21st century, the plastics were deteriorating. The plastic models were mostly clear but sometimes included solid colors, and some were a mixture of both translucent and opaque plastics. Realizing that the plastics were degrading, the curator enclosed some in polyethylene (PE) bags and stored them together in drawers within a closed wooden cabinet in an attempt to protect the other models. The metal handles and card holders on those drawers corroded, indicating that the segregated plastics were off-gassing acids that permeated through the PE bags.

Based on the results of monitoring with A-D Strips and Cresol Purple (Chapter 7), as well as testing with Diphenylamine and Nitrate spot tests papers (Chapter 3), combined with the age of the models and their often warped appearance, the majority of the models were determined to be CA with a few pieces of CN. And, in some cases, both plastics were used together.

After monitoring the individual plastics, A-D Strips were placed in the storage cabinet for one week to determine how acidic the general environment was from exposure to the degrading models; all test strips turned bright green, confirming an acidic environment. Recommendations were then made to the curator to remove the models from the PE bags and place them in a ventilated space, separate all the plastics models from the paper ones, and deaccession the heavily degraded plastics.

The options for housing the plastic models that would stay in the collection included placing them back in the closed wooden shelving units, putting them back in the wooden cabinets but keeping the doors off to increase ventilation, or placing them on open metal shelving to provide the most ventilation. Though not ideal to store plastics inside wooden enclosures, given the history of water leaks in the storage room, it was decided to keep the models in the cabinets but remove the doors. Drawers were rearranged for increased spacing, and tissue paper was hung in front to further limit dust and light while still permitting air exchange. Since the tissue paper will adsorb off-gassed acids, it needs to be periodically changed.

Five of the models were so severely degraded that the curator decided to deaccession them. Diphenylamine spot tests indicated that two of the five models were CN (Figure 9.3). The other three were clear CA with white CN internal forms; the combination of these two problem plastics likely accelerated the degradation of each other, as discussed in Chapter 4. The CA outer forms were yellowed, weeping, and warped (Figures 4.2 and 4.3).

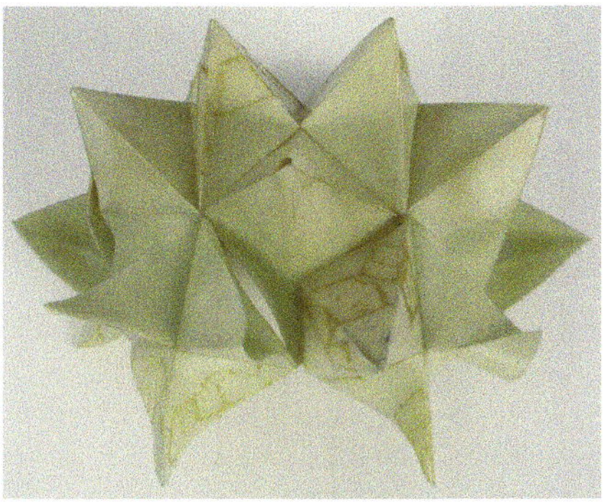

Figure 9.3 Circa 1930s geometric model made of a CN that is now deteriorated with scattered tape residue from previous repairs. Once in the NMAH mathematics collections (NMAH 1979.0102.254). Photographed by author, Division of Medicine and Science, National Museum of American History, Smithsonian Institution.

The CN model in Figure 9.3 was stored in a PE bag that, when opened, released an overpowering, menthol-like odor, likely caused by camphor plasticizer. The model has lost a lot of plasticizer, as evidenced by its brittleness, breakage, distortion, and odor. Discolored tape outlines, likely from previous repair attempts, are scattered along the edges. For these models, the distracting change in appearance and off-gassing of acids, as alerted by corroded metal hardware on the storage cabinets and confirmed with A-D Strips, informed the decision to deaccession.

Key Points for Plastics Collection Management Part 2: Discussions and Decisions

There are more question marks used in this chapter than anywhere else in the book, which is by design. Questions need to be asked to start discussions in order to reach decisions that become policies. The discussions that take place and the decisions that are made concerning accessioning, exhibiting, loaning, replacing, replicating, and deaccessioning plastics may look different at each institution. This is to be expected, as each museum has its own mission, may have different types of plastics in their holdings, and the stakeholders taking part in discussions and decision-making bring their own perspectives. That is why it is important to keep records of the rationale behind the decisions that are reached and become the protocols and procedures that are followed. Understanding why we are doing what we are doing is important, as is realizing that while setting precedent, past decisions do not have to be set forever. Rather, they should be periodically evaluated to ensure they are in-line with current practice and understanding.

References

Barone, Gabrielle. 2018. “Conserving Spok’s Ear.” *Inside the Conservation Lab* (blog). September 6, 2018. https://airandspace.si.edu/stories/editorial/conserving-spocks-ear

Coles, Fran. 2008. “Challenge of Materials? A New Approach to Collecting Modern Materials at the Science Museum, London.” In *Plastics Looking at the Future and Learning from the Past – Papers from the Conference Held at the Victoria and Albert Museum, London, 23–25 May 2007*, edited by Brenda Keneghan and Louise Egan, 125–131. London: Archetype Publications.

Conserv. 2024. “Understanding TWPI: The Time Weighted Preservation Index.” Conserv. https://support.conserv.io/knowledge/understanding-twpi

Elkin, Lisa and Waller, Robert. 2019. In *Preventive Conservation: Collection Storage*, edited by Lisa Elkin and Christopher A. Norris, 848–851. New York, NY: Society for the Preservation of Natural History Collections.

Floray, Steven P. 2004. “Disposal of Cellulose Nitrate Film.” Conserve O Gram 2/22. National Park Service. www.nps.gov/museum/publications/conserveogram/02-22.pdf

Guggenheim. 2022. “The Afterlife of Eva Hesse’s “Expanded Expansion”” (documentary). July 20, 2022. www.guggenheim.org/video/the-afterlife-of-eva-hesses-expanded-expansion

ICOM (International Council of Museums). 2017. “ICOM Code of Ethics for Museums.” https://icom.museum/wp-content/uploads/2018/07/ICOM-code-En-web.pdf

ICOM (International Council of Museums). 2019. “Guidelines on Deaccessioning of the International Council of Museums.” https://icom.museum/wp-content/uploads/2019/10/Guidelines-on-Deaccessioning-of-the-International-Council-of-Museums.pdf

Lodder, Christina. 2007. “Naum Gabo and the Quandaries of the Replica” *Tate Papers* no. 8 www.tate.org.uk/research/tate-papers/08/naum-gabo-and-the-quandaries-of-the-replica

Nishimura, Douglas, W. 2011. “Understanding Preservation Metrics.” Rochester, NY: Image Permanence Institute. https://s3.cad.rit.edu/ipi-assets/publications/understanding_preservation_metrics.pdf

NMAH (National Museum of American History). n.d. “Search Collections: Spock Ear Tip Worn by Leonard Nimoy in Star Trek II: The Wrath of Khan.” Smithsonian Institution. Accessed May 2, 2024. https://americanhistory.si.edu/collections/search?edan_q=star%20trek%20ear

Shashoua, Yvonne. 2008. *Conservation of Plastics: Materials Science, Degradation and Preservation.* Oxford: Elsevier/Butterworth-Heinemann.

Shpargel, Sara. 2003. “Environmental Management through the Climate Notebook™.” *Topics in Photographic Preservation* 10 (12): 108–125. https://resources.culturalheritage.org/pmgtopics/2003-volume-ten/10_12_Shpargel.html

Williams, Graham and Nina Williams. 2007. “Replicas of Constructions by Naum Gabo: A Statement by the Copyright Holders.” *Tate Papers* no. 8 www.tate.org.uk/research/tate-papers/08/replicas-of-constructions-by-naum-gabo-a-statement-by-the-copyright-holders

10 Collection Care Approaches for All Plastics

It can be overwhelming to see storage cabinets full of plastics or glance around an exhibition and realize just how many plastics there are. And if the decision is made to survey plastics in order to determine their possible types and current condition, as discussed in Chapter 8, starting the process can be intimidating. How do you begin to get an understanding of what is in the collection, what care should ideally look like, and make decisions around the potential impacts of plastics on other accessioned items? For a manageable approach for plastics care, these nine steps can be a helpful guide:

1. Observation
2. Confirm observation
3. Separate and isolate?
4. Clean area
5. Evaluate support
6. Evaluate environment
7. Document
8. Replace or replicate?
9. Deaccession?

1 Observation

The first step in assessing if any issues are present with plastics is observation. Look at the plastic itself. Is the plastic weeping, crizzled, cracked, sticky, smelly, warped, discolored, crumbling, has a bloom, or other change? If so, the plastic is deteriorating. Can the degradation be used to identify the likely type of plastic (Table 4.2)? Does that plastic pose any risks to other materials, such as through off-gassing (Chapter 4)?

During visual examination, some deterioration may be obvious but even in these cases sometimes the extent of degradation can be concealed. It can be helpful to examine plastics under high-intensity longwave ultraviolet (UV) light to help reveal the fuller extent of damage or change. In Figure 10.1, the presence of cracks

DOI: 10.4324/9781003489658-10

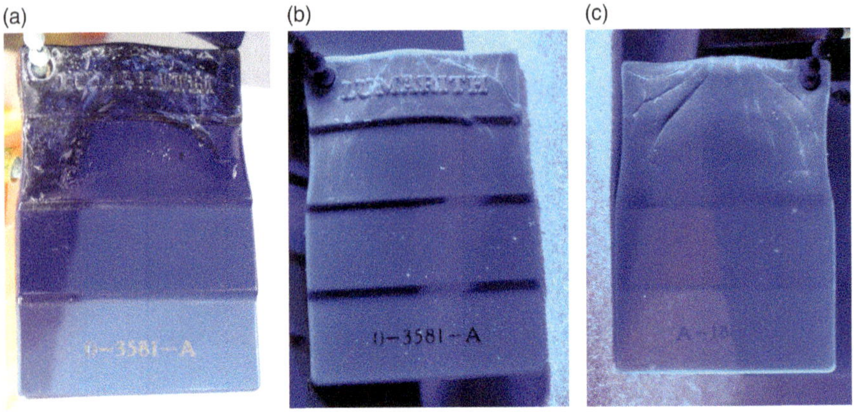

Figure 10.1 Lumarith salesman's kit sample coupon. (a) Front in visible light, (b) Front in UV light (365 nm), (c) Back in UV light (365 nm). Without UV examination, cracks were obscured by weeping materials. NMAH 1983.0538.61. Photographed by author, Division of Medicine and Science, National Museum of American History, Smithsonian Institution.

was obscured in visible light by material weeping from the cellulose acetate (CA) but the cracks were very apparent under UV light, thus revealing that the plastic was in a more fragile state than was first evident. In Figure 10.2, UV examination helped confirm that the cellulose nitrate (CN) sample, now degraded, originally would have mimicked tortoiseshell which aided in understanding the degree of change from its original form. In order to compare future changes, it is important to document the wavelength of UV light used in examination. The range for longwave UV light is from 320 to 400 nanometers (nm); the UV lamp used in Figures 10.1 and 10.2 was 365 nanometers.

Are degradation products being released from the plastic? Look at other objects near it. Are there any surprising deterioration or marked changes? If they are organics, are they discolored, perhaps darkened or yellowed, or embrittled? Is there a noticeable film or fogging on the inside of the case? Is there an odor? Are metals corroded? Be sure to look at accessioned metals as well as the hardware on the cases and cabinets like pulls, hinges, and screws. If these are corroded and the issues seem to be concentrated near the plastic object, it may be that it is either currently off-gassing acids or did so in the past (Sidebar 9.3). Remember that even if the plastic looks fine, if there are signs of deterioration nearby that could be caused by it, the plastic should be more closely evaluated.

2 Confirm Observation

If observation raised concerns that the plastic, or plastics, may be degrading, try to confirm it. Tools in Chapter 7 can be utilized to determine if degradation products

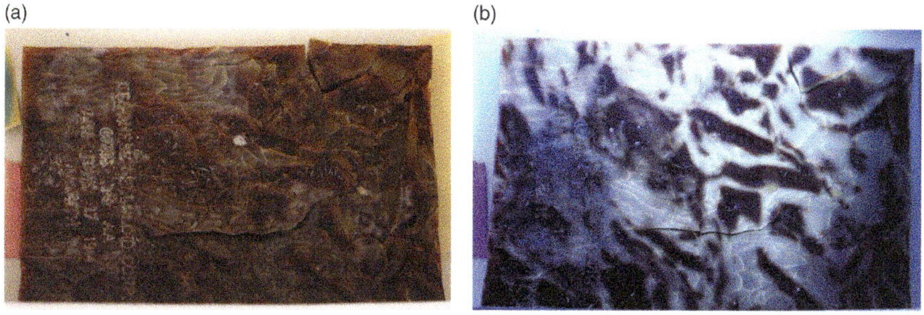

Figure 10.2 CN sample in (a) visible light, (b) UV light (365 nm). NMAH 1983.0538.83. Photographed by author, Division of Medicine and Science, National Museum of American History, Smithsonian Institution.

are being emitted that either are affecting nearby materials or have the potential to once deterioration progresses and emitted products get concentrated. Options for confirming the presence of degradation products released from plastics include the following:

- *Metal coupons* (silver, copper, lead). Long-term monitoring of all plastics.
- *A-D Strips* (turn from blue through shades of green to yellow in the presence of acids). Short-term monitoring of all plastics.
- *Cresol purple or red* (change from yellow to bright pink or reddish purple). Long-term monitoring of CN.
- *Ozone Test sticks* (change from white to yellowish beige to brown). Short-term monitoring of plasticized polyvinyl chloride (PVC).

Remember that other materials such as wood can release acids so it would be helpful to confirm whether the contamination is coming from plastics or something else. To counter the potential formation of an acidic soup, ideally plastics should never be stored in wooden enclosures.

3 Separate and Isolate?

When off-gassing or otherwise degrading plastics are found, the common reaction has often been to seal them in polyethylene (PE) bags thinking that will isolate the plastic and protect other materials. However, as covered in Chapter 4, concentrating degradation products can accelerate autocatalytic reactions of some plastics. If used as an enclosure, PE can adsorb the degradation products (Figure 3.1) and eventually the vapor can pass through it, as discussed in Sidebar 9.3. Additionally, PE can actively leach out plasticizers in PVC as described in Chapter 6. So, while well-intentioned, the decision to enclose plastics in PE can have a variety of negative consequences and should be avoided when possible, realizing that there may be exceptions because of health concerns (Sidebar 8.6).

If a plastic is a risk, it should be isolated. If the plastic is to remain in the collection, the platinum response is to place the plastic in low temperature or anoxic storage in an effort to slow the rate of degradation. This transfer can be a means of permanent storage or serve as a holding space until a decision about care or whether the plastic should be deaccessioned is reached. However, this route is likely out of reach for most plastics even in institutions that have cool, cold, or frozen storage because those resources are often limited and reserved for objects of high value or that pose a hazard, like CN film. Anoxic storage would slow oxidation degradation but can be difficult to maintain and is not utilized as frequently.

The middle level approach, which is still likely out of reach for many because of space constraints in storage, is to create a designated ventilated case for problem plastics. This segregation would permit easier monitoring and protect other collection materials from being affected. But by concentrating degrading plastics together, even if vented, the build-up of off-gassing could be detrimental to those plastics themselves. And having what amounts to a sacrificial case may be hard to justify.

The more accessible though less ideal approach for isolating degrading plastics could be to put them on open shelving and use Mylar/Melinex or even aluminum foil as a physical barrier if the shelves abut other shelving units. Ideally, monitor with A-D Strips to see how far any acidic off-gassing is traveling, avoid putting sensitive materials close by, and determine if there are health risks to people working nearby.

Regardless of the option available, there is value in separating an actively degrading plastic from other materials until a decision about longer-term care or deaccessioning is reached.

4 Clean Area

When degrading plastics are found, remember that their housings including boxes, trays and the cabinet, case, or shelf itself are contaminated. Once the immediate housing like a tray is properly discarded, decide how to clean the permanent fixture. Wiping down the shelves or drawers with water could help neutralize some deposition. Then, if the case or cabinet was sealed, is it possible to vent? This may mean leaving the doors open or taking them off. The vents that are often built into the doors of museum quality metal cabinets may not provide enough air exchange if the accumulation of degradation products was very high but if that is the only option it could still help, just at a slower rate.

If ventilation is not possible, perhaps because of concerns of contaminating nearby materials or because of a strong odor that may affect staff working in the room, scavengers like zeolites or activated carbon/charcoal could be sealed in the case in order to adsorb the remnants of the released pollutants. Of the activated carbon/charcoal products available, carbon/charcoal cloth was found to scavenge the most acetic acid from off-gassing CA when compared to activated carbon/

charcoal foam or loose granulates and had the added benefit of not generating dust or leaving carbon black particles behind (Grøntoft, Lankester, and Thickett 2015). As a reminder, these sorbents can be too aggressive for some plastics and may pull out plasticizers so use with caution if any plastics remain in the case or cabinet (Chapter 4). For some museums such as the Tate, the inclusion of activated carbon/charcoal cloth in storage enclosures for CA artworks is routinely employed, monitored, and assessed as a preventive measure (Townsend, Hackney, and Kearney 2023).

After cleaning, venting, or scavenging, place A-D Strips in the sealed enclosure and monitor for a week. If the strips react, residual acid is still present and the process of cleaning then venting or adsorbing should be tried again, perhaps for a longer period of time. Or if venting was attempted, change to scavengers. Or if adsorbents were not enough, try venting. Once the reaction of the A-D Strips indicates that the enclosure is safe, move stable collections back in.

5 Evaluate Support

Storage trays, boxes, mounts, or other supports that were exposed to degrading plastics need to be replaced for the sake of both the plastic itself and any nearby objects. These supports should never be reused. As seen in Chapter 4, materials like paper and plastic can adsorb volatile organic compounds (VOC) and acidic vapor. This not only causes them to discolor and become brittle but they can in turn release the adsorbed contaminants, negatively affecting material stored in them or near to them. You will ultimately spend more money redeploying these supports and damaging collections than you would save from reusing them.

If new supports are needed, it is a good opportunity to evaluate storage material choices and their potential effects on plastics, as covered in Chapter 6. Remember that buffered archival corrugated boards, like blueboard, are generally safe or outright beneficial for use with plastics, as covered in Chapters 4 and 6. For some plastics, adding silicone release Mylar/Melinex at points of contact (Sidebar 10.1) or silicone release paper (Sidebar 4.1) may be prudent. Since some plastics may become less flexible over time, use this opportunity to design a housing that will more fully support the plastic in a position that will retain as much of its original form as possible.

6 Evaluate Environment

Once steps 1–5 are complete, look at the broader environment to determine the potential role(s) light, temperature, relative humidity (RH), and pollutants may have played in plastic degradation, as discussed in Chapter 5. Remember that light is damaging to many plastics. Eliminating UV light exposure, storing in the dark, and exhibiting at low visible light levels can all help slow the rate of light-induced degradation.

High temperatures increase the rate of all chemical reactions; thus, lowering temperature can have a significant impact on prolonging the lifetime of plastics. Achieving cool, cold, or frozen temperatures requires a lot of resources and is not feasible for many collections. That said, any reduction of temperature can be beneficial. So, if it is possible to have a storage room at about 19°C (66°F) instead of around 22°C (72°F) that could help plastics a bit without too much discomfort to staff. Of course, as described in Chapter 5, lower temperatures can raise RH, which comes with its own risk for some types of plastics (Chapter 4). Additionally, if the heating or cooling system is fighting the outside environment too much, it can cause strain and may result in condensation issues either internally or externally. So, decisions about environmental regulation need to consider the unique variables imposed on the collections by a building with its set of environmental controls, which may be aged, and what can be achieved at a specific location at various times of year. To reach a decision, conversations between collection staff and facilities maintenance are vitally important.

Plastics, like most materials, benefit from a stable RH with minimal fluctuations generally in the 45–55% RH range with an allowable drift of +/−5%, which works out to an RH range of 40–60% over the course of year. But as outlined in Chapters 4 and 5, there are exceptions and RH that is too high or too low can be damaging to some plastics. Since collections are often stored mixed together, the targeted RH set point and range may be set as a compromise between needs or determined by the plastics that are most sensitive, most valuable, or most prevalent (Plastic Identification Tool n.d.).

When viewed through the lens of this book, it is assumed that the main pollutant source in a collection is or may be a plastic. Remember though that other materials, whether in the collection or used to house it such as a case or cabinet, may also introduce reactive pollutants.

7 Document

If collections are moved for monitoring or because a contaminated case or cabinet needs to be cleaned, changes to location obviously need to be tracked. Observed changes in condition of the plastic should also be documented, as outlined in Chapter 8. By recording in written and photographic forms, the degree of degradation can be compared to past records which may give a sense of how quickly deterioration has developed or how far it has progressed over a period of time. This information can then be used in any subsequent conversations regarding insurance and potential deaccessioning. And if the plastic is deaccessioned, the documentation will be all that remains of the plastic in the collection.

8 Replace or Replicate?

Does the museum have a policy around replacement or replication? If so, is the degrading plastic a candidate for either? If no policy exists, could the plastic of concern spark those conversations? Who are the stakeholders who should be involved both generally and for the specific plastic in question? If conversations on replacement or replication occur, they need to be documented and archived, as discussed in Chapter 9.

9 Deaccession?

Evaluation of existing documentation compared to the current condition should help guide whether the option of deaccessioning, the last step in the process of how to approach plastics, is pursued. In the context of this book, the decision to deaccession is driven by poor condition which may pose a threat to other materials in the collection or cause the museum to be unable to care for the object, as covered in Chapter 9.

Sidebar 10.1 Plastics in Julia Child's Kitchen

For over 20 years, the installation of Julia Child's kitchen in the Smithsonian's National Museum of American History (NMAH) has been a popular attraction for visitors. In 2011, ten years after the kitchen was acquired, Museum Studies graduate students at The George Washington University (GW) conducted a condition survey under the tutelage of two NMAH curators, Paula Johnson and Steve Velasquez, and GW professor, Mary Coughlin, as part of preparations to move the kitchen across the building into a new exhibition space.

As a kitchen that was in use from 1961 to 2001, plenty of plastics are present including PVC and silicone (SI) as well as unidentified rigid plastics used in various kitchen tools. The class essentially approached the plastics using the nine steps outlined in this chapter:

1. Observation
2. Confirm observation
3. Separate and isolate?
4. Clean area
5. Evaluate support
6. Evaluate environment
7. Document
8. Replace or replicate?
9. Deaccession?

During the semester, the five students in the class surveyed about 400 objects and the first step was always observation. Obvious condition issues for plastics included a weeping SI spatula, tacky PVC-covered tablecloth, weeping PVC dishrack, a slightly tacky phonebook cover likely made of either PVC or polyurethane (PUR), and a pair of discolored Bluettes brand dishwashing gloves with a chloroprene (better known by the brand name Neoprene) outer shell.

In order to confirm if these plastics posed a threat to nearby materials, students set out A-D Strips for one week. The dishrack and the dishwashing gloves turned the strips partially or totally yellow, indicating high acid concentrations. The PVC-covered tablecloth, phonebook, and SI spatula did not register as off-gassing acids at the time of testing.

Though protocol may be to separate and isolate off-gassing plastics, what would taking the dishrack and gloves off display do to the interpretation of the kitchen? Should the spaces be left empty? Could they be replaced with reproductions or similar looking items? At what point would the kitchen no longer be in its original state as used by Julia Child and how would that impact the curators' goals of keeping the kitchen as original as possible? But what risks would the degrading plastics pose to other objects if they remained on display? Would those risks increase over time? One student's view on the discolored, off-gassing dishwashing gloves was to "...definitely recommend considering removing the gloves from the kitchen." While at first this seems like reluctance to make a recommendation, it actually captures how nuanced collection care choices can be when it comes to plastics and the need to have multiple stakeholders involved in decision-making.

Rather than removing the degrading plastics, the decision was made to deploy another round of A-D Strips to see how far the acidic vapor was spreading and if it was impacting nearby materials. The off-gassing did not extend far from the sources so the decision was made to leave the dishrack and gloves on view.

Whether off-gassing or not, it was evident that the weeping and discolored plastics were already on their degradation pathways, but since they would remain on display, conservators cleaned off weeping materials. The students then added silicone release Mylar (Melinex) at points of contact between currently weeping plastics and the material supporting them: at the interior bottom of the ceramic crocks that contain the plastic spatulas, beneath the phonebook on the wooden butcher-block countertop, under the placemats and bowl on the tablecloth, below each object in the drying rack, and under each dishwashing glove. In order to protect against potential future negative interactions between other plastics and the wooden countertop, silicone release Mylar (Melinex) was also placed underneath the coffee maker, lazy Susan, cutting board, feet of the KitchenAid mixer, and the tray holding the dishwashing gloves.

These measures proved prudent. Ten years after the class, the SI spatula completely dried out (Figure 10.3); the dishrack darkened and was

(a)

(b)

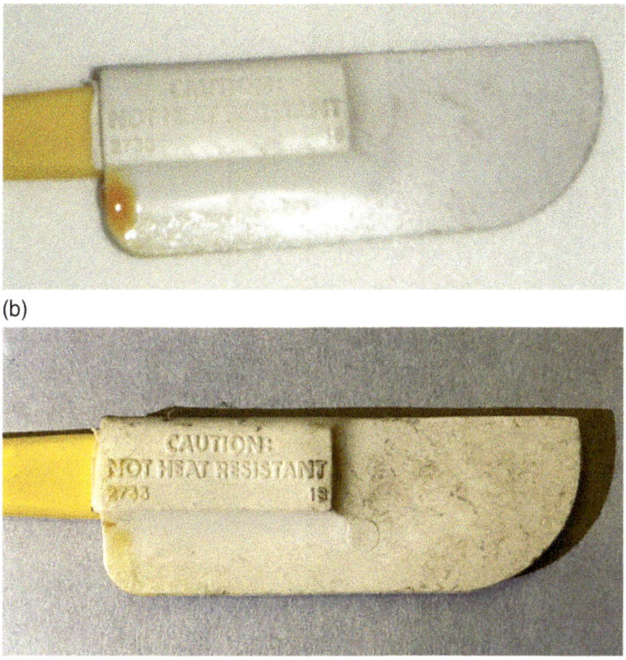

Figure 10.3 (a) Silicone rubber spatula in 2011; orange weeping material pooled at bottom and dust stuck to the surface. (b) Spatula in 2021; orange weeping material dried, surface darkened, and dust stuck to surface. NMAH 2001.0253.0707. Photographed by author, Division of Home and Community Life, National Museum of American History, Smithsonian Institution.

embrittled, presumably from the loss of plasticizers; and the dishwashing gloves markedly changed: white crystals formed, likely from plasticizers that crystallized on the surface, the seams of the bottom glove separated because of stress from the weight of the glove on top of it, a few fingertips appeared to have shrunk, and the top dishwashing glove turned from blue to black. The color change was likely catalyzed by light exposure since the blue color of the bottom glove was relatively unchanged (Figure 10.4). Each glove stuck to the interleaving, but because it was silicone release Mylar (Melinex), it could be separated without damage. Without that layer, the gloves would have stuck to each other and the tray they were placed in, likely resulting in physical damage. An added benefit of the interleaving is that each glove can be lifted up and evenly supported, enabling handling without touching the gloves.

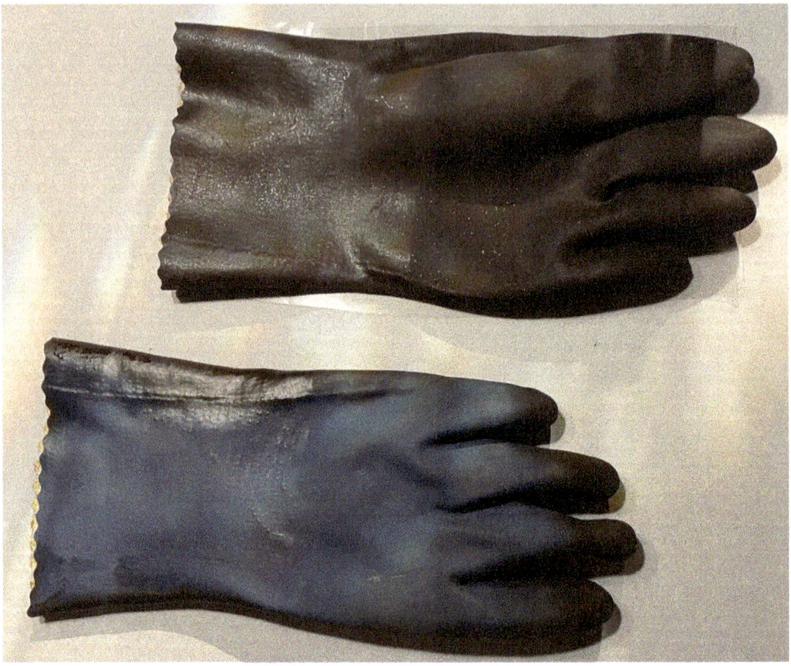

Figure 10.4 Dishwashing gloves in 2021. Top black glove was blue in 2011 but turned completely black ten years later. The bottom glove was protected and retained most of its blue color. Photographed by author, Division of Home and Community Life, National Museum of American History, Smithsonian Institution.

Exhibitions are often a compromise between accessibility and preservation. As discussed in Chapter 9, navigating this balance is particularly tricky with some plastics. The kitchen is an immensely popular exhibit that has been enjoyed by tens of millions of visitors over the last two decades, which far exceeded original expectations for the display duration. It is true that some of the plastics, like the dishwashing gloves, would have fared better if kept in dark, low temperature storage but doing so would not help fulfill the museum's vision to "… challenge ourselves to become the country's most accessible, inclusive, relevant, and sustainable public history institution" (National Museum of American History 2024). Therefore, NMAH staff continue planning for the protection of the objects individually and collectively while being committed to having them remain accessible.

The continued inclusion of some degrading plastics in the kitchen seems to be having minimal impact on other materials. However, if that changes and the plastics need to be removed from display and then pose a risk in storage, conversations about isolating or deaccessioning may be warranted. The need to periodically monitor plastics for changes in appearance, such as increased weeping, formation of crystals, discoloration, deformation from loss of mass, formation of cracks, crumbling, or total breakdown of the form, is important because at this stage the plastic is likely unable to be exhibited or doing so may require explanatory label text. As covered in Chapter 7, it is important to monitor plastics for off-gassing that may be spreading out and affecting other collection materials. Remember, periodic monitoring is important because in some cases the release of gaseous degradation products can occur before visible signs of deterioration are evident.

As students who were required to take two classes in preventive conservation in order to graduate, the class knew the impact the environment can have on condition, so they monitored light, temperature, RH, and evaluated dust as a pollutant.

Students monitored visible light levels throughout the kitchen and tracked results in an Excel spreadsheet noting whether the visible light levels were appropriate or inappropriate. Because of the types of bulbs and the interior location of the exhibit, UV light was not an issue. Light levels were taken again two years after the class when the kitchen was exhibited in its new location and improvements were evident. Visible light levels around the phonebook decreased from between 38 and 140 lux down to a range of 14–26 lux, which is well below the 50 lux maximum for plastics. Mounted to the wall on either side of the sink is a pair of lights that originally, and at the time of the class, had incandescent bulbs pointing downward, resulting in hot spots of both light and temperature on the materials below. The objects most at risk included the dishrack and dishwashing gloves with levels ranging from 211 to 430 lux depending on the intensity of the light in a location. Since the class, the incandescent bulbs have been replaced with LEDs that mimic the look and warm color temperature of the original bulbs but with a reduction in lux levels and heat. LEDs also have the benefit of using less energy and have much longer timespans between bulb replacement (the current bulbs have been in place for at least four years). Thus, using LEDs has reduced energy usage, lowered costs for bulb replacement, and required less staff time to monitor the bulbs and replace them, leaving more time for other exhibit maintenance tasks.

A digital hygro-thermometer was placed in the kitchen to record the temperature and RH at the start and end of each 2-hour class. Generally, temperature readings increased by about 1.1°C (2°F) over each month from a low of 22.1°C (71.8°F) in February to 24.6°C (76.4°F) in April. The temperatures were likely artificially raised because of seven to eight people working in a relatively small space but were still higher than recommended upon entering the kitchen. Lowering the temperature would slow the rate of chemical degradation but getting it low enough to really make an impact on the plastics is not possible in an exhibition. The RH reached a low of 36% in February, likely because of the inverse relationship to heating, and leveled out in March and April staying between 42 and 45%, which is within the general recommendations for plastics. In the new exhibition space occupied since 2012, the environment remains within recommended ranges for most plastics (Chapter 5).

The original exhibition space from 2002 to 2011 was meant to be open for a much shorter time period and had an open ceiling which allowed a lot of dust to deposit and stick to weeping plastics, as seen in Figure 10.3. NMAH staff were aware of the issue and cleaned the kitchen as often as possible. The students noted concerns of dust deposition in the condition survey and the NMAH curators and conservators fought to include a ceiling in the new exhibition, which has greatly reduced dust levels.

The documentation generated from the time of accessioning and installing into the first exhibition, the class survey 10 years later, conservation treatments, and the status of the plastics 20 years after coming to NMAH craft a story of how the plastics are changing over time. Plastics require periodic conversations about care because the rate of change is often quicker than other materials which means we notice changes within the span of our careers or, in some cases, after just a few years. The decisions that are reached reflect views at a moment in time and may change. At the opening of the first exhibit in 2002 and at the time of the class in 2011, the curators were dedicated to authenticity of the kitchen defined as everything in the kitchen being what Julia used and donated (besides the paper floor covering made to match the original commercial-grade vinyl flooring and the imitation bananas and tomatoes placed in baskets to suggest foods Julia typically had on her table and countertops). Over the last few years, however, the curatorial view on authenticity has begun to shift as some plastics are naturally aging out of their exhibitable lifetimes. Whether or not some of the plastics are removed, replicated, replaced, or deaccessioned will be part of future conversations that will likely need to periodically occur, as discussed in Chapter 9.

Key Points for Collection Care Approaches for All Plastics

Despite the potential risks associated with plastics, having them in your collections does not have to lead to certain doom or even eventual replacement, replication, or deaccession, though in some cases those are prudent decisions. Through observation, examination, and monitoring, deteriorating plastics can be identified and measures taken to protect other materials. Ventilation, the potential addition of scavengers, rehousing, and the creation of as appropriate an environment as possible are all tools that can be deployed. With these relatively simple steps, responsible stewardship of plastics is possible regardless of the size of the institution, the number of staff, or the budget allocated for collection care, making caring for plastics in collections an achievable goal for all.

References

Grøntoft, Terje, Paul Lankester, and David Thickett. 2015. "Reduction of Acidic Pollutant Gases Inside Showcases by the Use of Activated Carbon." *e-Preservation Science* 12: 28–37. www.english-heritage.org.uk/siteassets/home/learn/conservation/collections-advice--guidance/reduction-of-acidic-pollutant-gases-inside-showcases.pdf

National Museum of American History. 2024. "About." Smithsonian Institution. Accessed May 31, 2024. https://americanhistory.si.edu/about

Plastic Identification Tool. n.d. "Plastics." Cultural Heritage Agency: Ministry of Education, Culture, and Science. Accessed May 31, 2024. https://plasticidentificationtool.nl/

Townsend, Joyce H., Stephen Hackney, and Mark Kearney. 2023. "A Critical Assessment of A-D Strips® Used in Degradation Studies of Artworks Made from Cellulose Acetate." *Studies in Conservation*, 68 (1): 9–17. https://doi.org/10.1080/00393630.2021.1996091

Index

Note: Page numbers in *italics* refer to figures and those in **bold** refer to tables.

9/11 Memorial & Museum, collections survey planning at 146–150

access, decisions on 158–159
accessioning 155–158; considerations for accessioned plastics 107–108; expectations towards 156; lifetime of plasticized PVC 155; problem plastics 155–157; Time Weighted Preservation Index (TWPI) 157–158
acetate 7, **41** *see also* cellulose acetate (CA)
acetic acid 21, 53, 54, **65**, 88, 89, 97–98, **116**, 117, 121–122, 123
acetone, labeling and 150
Acid-Detecting (A-D) Strips 54, 59, 89, 90, 117–122, 123, *124*, 141, 171, 172, 173, 176; deaccessioning mathematical models 166–167; general testing protocols for 120–121; monitoring cellulose acetate necklace 121–122; packaging and storage of 121; reference card 118–119; use with Ozone Test strips 125–128
acrylic **7**, 16, **41**, 42, *43*, 80, **95**, 101 *see also* polymethyl methacrylate (PMMA)
activated carbon/charcoal 51, 54, 90, 173
additives 16–24; bisphenol A (BPA) 19, 58, 96; coloring and 21–23; di-(2-ethylhexyl) phthalate (DEHP) 57–58; health and safety aspects 17–20; polyhydroxyalkanoate (PHA) 24; textiles 17; triphenyl phosphate (TPP) 20–21, 50, *84*, 85
Ageless (oxygen scavenger) 90
aging phenomena: atlas of 143–144; monitoring 142
American Institute for Conservation (AIC) 109–110, 111, 115
anoxic storage 90–91, 172
antioxidants 16, 21, 96, 99
antistatic agents 16
archeological research, storage and sample viability 104–106
artist interview 132–134; Shervone Neckles 161–164
Attenuated Total Reflectance Fourier Transform Infrared Spectroscopy: (ATR-FTIR) 30, 142

Backer rod 96, 97
Bakelite (phenol formaldehyde) **13**, 14, 42
Beilstein Test 30, 58, 61
biobased plastics 9; environmental impact, 24
biodegradable plastics 9; environmental impact 24
birefringence 39–41
bisphenol A (BPA) 19, 58, 96
black plastics, storage supports 94
blowing agents 16
Blue Wool Standard 71–72
blueboard 51, 54, 66, 82, 90, 107, 173
Bubble Wrap 98
burn tests 29–30
buttons 55–56

camphor 48, 50; smell 65, **144**, *167*
Celluloid **13**, 20, *29*, *34*, 35, 48–49, *117*
Celluloid Corporation 20, 34–35
cellulose 7, 12, 48, 53
cellulose acetate (CA): background 52–53; buttons 55–56; deterioration 53, *54*, **65**; eyeglasses monitoring for CA/CN

123–124; monitoring necklace 121–122; preventive conservation 54; RH range 80
cellulose nitrate (CN): background 48–49; buttons 55–56; deterioration 49–50, *51*, **65**; eyeglasses monitoring for CA/CN, 123–124; fire risk of waste 165; preventive conservation 50–52; reformatting film 165; RH range 80
cellulose triacetate (CTA) 80, *84*, 86–88
Choosing Materials for Storage, Exhibition & Transport (wiki) 109–110
CIELab 72–73
cleaning 172–173
"cleanroom" PE plastics 100
collection care 169–181; cleaning 172–173; documentation 174; environment evaluation 173–174; observation 169–171; replacement or replication 175; separation and isolation 172; supports 173; ventilation 172
collection management decisions: accessioning 155–158; deaccessioning 164–167; exhibition and loans 158–159; replacement or replication 160–164
collection management processes and procedures: databases 138–150; documentation 132–138; handling 130–132; labeling 150–152
collection records and timelines 41–44
color, recording on documentation 135
color change, identifying and tracking 135–136
colorants 16
coloring, additives and 21–23, *24*
composite objects, environments for 91
Conservation and Art Materials Encyclopedia Online (CAMEO) 108–109
Construction in Space: Two Cones (copy artwork) 160
Cooper Hewitt, Smithsonian Design Museum 60, 63; plastics collection survey 140–142
Coroplast **95**, 100
Correx **95**, 100
Costume Institute, Metropolitan Museum of Art 55–56, 76–77, 90–91, 121–122, 135–136
cotton gloves 131–132
Cresol Purple/Red Indicator Strips 122–125, 128, 171
critical moisture indicator (CMI) packaging method 77
Curator's Guide to Plastics (website) 36
Curator's Guide to Synthetic Garments 37

3D printing 9–11; yellowing of printed surface 135–136
data loggers: relative humidity 81–82; temperature 75–76
databases: aging phenomena 144–145; thesauri 138–139
deaccessioning 164–167, 175; documentation of 174; mathematical models 166–167
degradation: 9/11 Memorial & Museum items 148–150; common signs of 46, **65**; terms used for 46, **47**
dehumidifiers 82
dehydrochlorination 57, 58, 107–108
Design Museum Gent, database thesauri 138–139
deteriorating plastic, Time Weighted Preservation Index (TWPI) 157–158
di-(2-ethylhexyl) phthalate (DEHP) 57–58
diffusion tubes 88–89
documentation 132–138, 174; artist interview 132–134; contents of 134–137; deaccession decisions and 132; identifying marks 134–135; Julia Child's kitchen 180; method of entry 134; photography use 138
Domiciliation: Bless This House, interview with artist 161–164
dosimeters 88–89
dust: Julia Child's kitchen 180; World Trade Center 148–149

Ebonite **14**, 46
elastomers (elastic polymers) **14**, 15
environment evaluation 173–174
equilibrium moisture content (EMC) 82, 83–85
Ethafoam **13**, 16, 94, **95**, 96–97, 103
ethanol, labeling and 150–151
exhibitions and loans, discussions and decisions on 158–159
extruded film 8–9
eyeglasses, monitoring for CA/CN 123–124

fillers 16
film storage 51–52
fire retardants 17; triphenyl phosphate (TPP) 20–21
phosphate (TPP)

fire risks: CN film 49, 51–52; material choices 106–107
fluorescent lights 70, 74
foams, PUR 62–63; deterioration 63–64
foot-candle 71
formic acid 88, 89, 123
Fourier transform infrared (FTIR) spectroscopy 30–31, 104, 126–127; ATR-FTIR 30, 142; PUR condition 64; PVC condition 58; specular reflection-FTIR 30
fused deposition modeling (FDM) 10, *11*

Gabo, Naum 160
gas chromatography/mass spectrometry (GC/MS) 30, 31, 32, 89, 104, 105; Py-GC/MS 31–32
glass transition temperature (Tg) 38, 52, 74
gloves 19–20, 131–132; dishwashing 176–178; nitrile gloves and metal coupons 114–115
Gore-Tex **13**, **95**, 103

handling 130–132; health impacts 17–20; human-made fiber-based textiles 130–131
hard rubber (HR) **14**; background 46–48; deterioration 48, **65**; preventive conservation 48
Harvard Art Museums 59, 77–79
hazardous materials: 9/11 Memorial & Museum 148–150; disposal of 165
headspace-solid-phase microextraction-gas chromatography/mass spectrometry (HS-SPME-GC/MS) 30, 32, 89; PUR condition 64
health risks (health and safety): 9/11 Memorial & Museum 148–150; handling plastics 17–20, 130–132
heat stabilizers 17
heating, ventilation, and air conditioning (HVAC) units 75, 76, 79, 81, 82–83, 89, 164
Hesse, Eva 158–159, 160
Hollytex **95**, 97, 103
human-made fibers **7**, 17
humidifiers 82–83
humidity expansion coefficient (HEC) 85–88
humidity indicator cards 81–82
hygroscopic materials 77, 82, 83–84, 85, 86; polyamide (PA) 95
hygrothermographs 75

identification *see* non-analytical identification; scientific analysis
Image file choice, documentation 138
Image Permanence Institute 42, 52, 54, 77, 137, 157; Acid-Detecting (A-D) Strips 117–122; Time Weighted Preservation Index (TWPI) 157–158
inflatables 107–108
infrared (IR) light **70**, 71; FTIR spectroscopy 30–31, 104, 126–127
insurance claims, importance of documentation 132
International Council of Museums (ICOM), deaccessioning guidelines 164–165
interviews *see* artist interview
isolation of degrading plastics 172

Julia Child's Kitchen 175–180

Kevlar **7**, **13**, 107
Kimdura 8
"Know, Name and Assess your Plastics" project 138–139

labeling 150–152; solvents 150–151; textiles 151–152
latex gloves 131
LDPE *see* polyethylene (low density) (LDPE)
light: CIELab 72–73; color monitors 72–73; do-it-yourself monitor 72; evaluation of environment 173–174; foot-candle 71; Julia Child's Kitchen 180; L*a*b* 72–73; lumen 71, lux 71; mitigation 73–74; monitors 71–72; plastic sensitivity to 69, **70**; principle of reciprocity for exposure 73; and PVC degradation 59; sensitivity of A-D strips 119; sources 69–71; spectrophotometer 72–73; types of 69, **70**
light-emitting diodes (LEDs) 71, 179
loans: discussions and decisions on 159–160; Spock ear tips loan request 159
low-density polyethylene (LDPE) **13**, 16, 33, 34, **40**, *43*, 77, **95**, 98–100, *99*
low-temperature storage 76–77; PVC low temperature storage 77–79
lubricants 17
Lucite **13**, 16, **95**, 101
Lumarith **13**, 20, *21*, 22, 34–35, 52, *170*
lumen 71
lux 71

3M Art Fabric 8
MarvelSeal 98, 121
material choices, resources to aid 108–110; online resources 108–110; Safety Data Sheets (SDS) 108
material safety data sheets (MSDS) *see* safety data sheets (SDS)
Materials Testing Results (wiki) 110
Materials Working Group (MWG) 109
measurements, recording on documentation 135
metal coupons 114–117, 128, 171; sensitivity **116**; using nitrile gloves with 114–115
Minimum Efficiency Reporting Value (MERV) 89
monitoring: Acid-Detecting (A-D) Strips 117–122; Cresol Purple/Red Indicator Strips 122–125; Julia Child's kitchen 179–180; metal coupons 114–117; Ozone Test Sticks 125–128
monitors: light 71–72; relative humidity (RH) 81–82; temperature 75–76
Museum of Design in Plastics (MoDiP) 36, 41, 42, *43*, 44, 48, 52, 75, 80
Mylar/Melinex **13**, 59, 65, *78*, 79, **95**, 101, 103, 107, 120, 121, 135, 172, 173, 176, 177

National Museum of American History (NMAH) *21*, *22*, *29*, *34*, *36*, *51*, *53*, *54*, *60*, *81*, *116*, *117*, *123*, *124*, *150*, *170*, *171*; deaccessioning mathematical models 166–167; Julia Child's kitchen 175–180; Spock ear tips request 159
natural plastics 6–7 *see also* individual plastics
near-infrared reflectance (NIR) spectroscopy 32–33
Neckles, Shervone 161–164
neoprene (chloroprene) **14**, *43*, 176; low-temperature storage 77
Nitrate Test papers 28–29
nitric acid 48, 49–50, **65**, **116**, *116*, 122, 123–124
nitrile gloves 19, 28, 46, 58, 114–115, 120, 122, 125, 131, 132
Nomex **7**, 107
non-analytical identification 33–44; collection records and timelines 42–44; online tools 36–41; polarizing filters and birefringence 39–41; resin identification code 33–34; trade names and patents 34–36
non-breathable microclimates, analyses affected by 106
Nylon (polyamide (PA)) **7**, **13**, **41**, *43*, 77, 95–96

observation, collection care 169–171
Oddy Test 89, 94, 103, 110, 114, 115, 116
odor/smell 18, 47, 50, 53, 58, **64**, 88, 96–97, 121–122, 131, 143, **144**, 148, 157, 167, 169, 170, 172
off-gassing 55, 59, 79, 83, 98, 121, 123; A-D strips and 117–120; CN degradation 50–51; deaccessioning and 166–167; health concerns and 131; isolation of degrading plastics 172; Julia Child's kitchen 176; PET as blocker 101; polystyrene 102; separate and isolate 172, 176; sources of 88
online tools, used for identification 36–41
organic residue analysis 105
oxidation 90, 102; LDPE 99; photo- 21, 41, 48, 63, 64, 69, 73, 95; thermal- 74, 98
oxygen scavengers 90
Ozone Test sticks 58, 59, 125–128, 171

pencil, used for labeling 151
personal protective equipment (PPE) 19–20, 28, 115
Perspex **13**, 16, 80, **95**, 101
Philadelphia Museum of Art 160
photochemical reactions 69
physical effects, analyses affected by 106
Plastazote **13**, **95**, 96, 97
Plastic Identification Tool 37–41, 42, 91
plasticity 12–16, 25
plasticized PVC **13**, **14**, *43*, 57, 58, 59, **70**, 101, 107, 125, 126, 127, 155; low-temperature storage 77–79; storage impact on watches 59–62
plasticizers 17, **47**; adsorbents and 51; CA and 53, 54; camphor as 48, 50, *167*; health risks 17–20, 131–132; PVC and 56–59, 77–79, 101, 107, 127, 171; yellowing and 59–62
Plexiglas **13**, 16, 80, **95**, 101, 160
polarizing filters 39–41, **40**
pollutants 88–90, 173; mitigation 89–90; monitors 88–89; sources 88
polyamide (PA) 10, **13**, *43*, **70**, 77, **95**, 95–96, 135–136
polycarbonate (PC) **13**, **40**, *43*, 76, **95**, 96, 150
polyester **7**, **13**, 24, **41**, 58–59, 64, 70, 77, 80, 90, 97, 101, 103, 131

polyethylene (high density) (HDPE) **95**, 96–98; fibers 97; foams 96–97; sheets 97–98
polyethylene (low density) (LDPE) **13**, 16, 33, 34, **40**, *43*, 77, **95**, 98–100; containers 98; film 98; sheeting 99; sleeves and bags 99–100
polyethylene (PE) 9, 12, 16, 28, 34, 35, **40**, 58, **70**, 74, 96, 166–167, 171; fire risk of sheeting 107 *see also* polyethylene (high density) (HDPE); polyethylene (low density) (LDPE)
Polyethylene – Polypropylene (PP) copolymer **95**, 100
polyethylene terephthalate (PET or PETE) 9, **13**, 33, **40**, 58–59, **95**, 101, 150
Polyflute **95**, 100
polyhydroxyalkanoate (PHA) 24
polylactic acid (PLA) 9, **13**
polymers 11–12
polymethyl methacrylate (PMMA) **13**, 16, 42, *43*, 80, **95**, 101–102, 150, 151
polypropylene (PP) 9, **13**, 33, *39*, **40**, *43*, **70**, 74, 98, 102
polystyrene (PS) 12, **13**, 16, 34, *43*, **70**, 77, **95**, 102, 150, 151
polytetrafluoroethylene (PTFE) **13**, *43*, **95**, 103
polyurethane (PU or PUR) 7; additives 64; anoxic storage 90–91; background 62–63; deterioration 63–64, **65**; ester/ ether 63, 64–65; fibers 64; low-temperature storage 77; preventive conservation 64–65; RH range 80
polyvinyl acetate (PVAc) **13**, 57
Polyvinyl Chloride (PVC) 7, 30; background 56–57; description in databases 139; deterioration 57–58, **65**; hazardous waste disposal 165; inflatables 107–108; lifetime of 155; not to be stored in PE 100; Ozone Test sticks and 125–128; preventive conservation 58–59; rehousing Joseph Beuys *Postcards* 77–79; Swatch watch storage 59–62, *60*; used with PET 58–59, 101
problem plastics 46–66; 9/ 11 Memorial & Museum 149; accessioning 155–157
processing aids 17
pyrolysis-gas chromatography/mass spectrometry (Py-GC/MS) 31–32; PUR condition 64
pyrolysis 31–32

radiocarbon dating 105
Raman spectroscopy 31; PVC condition 58
rayon 7, **41**, 52
relative humidity (RH) 79–88, 173; A-D strips and 119; CA degradation and 53, 54; CN degradation and 49–50, 51–52; evaluation of environment 173–174; fluctuations 80–81; Julia Child's kitchen 179–180; mitigation 82–88; monitors 81–82; related to temperature 74–75, 79–80, 82; sources 80–81
replacement or replication 160–164, 175
replicas 160
resin identification code 15, 33–34, 98, 102, 134
rubber, natural 6–7

Safety Data Sheets (SDS) 20, 108
sample viability, storage and archeological research, 104–106
scanning electron microscopy/energy-dispersive spectrometry (SEM/ EDS or EDX) 104
Science Museum (London), accessioning problem plastics 156–157, 165
scientific analysis 27–33; burn tests 29–30; FTIR spectroscopy 30–31; HS-SPME-GC/MS 32; instrumental techniques 30–33; NIR spectroscopy 32–33; Py-GC/ MS 31–32; Raman spectroscopy 31; spot tests 27–29; XRF 32
selective laser sintering (SLS) 10
semi-crystalline/amorphous 16
semi-synthetic plastics 7 *see also* individual plastics
separation of degrading plastics 171
silica gel 83
silicone (SI) **14**, *43*, 77, 90, **95**, 101, 103, 175, 176, *177*; release paper 56, 103, 107, 108, 135, 173; *see also* Mylar/ Melinex
solubility parameters 150–151
solvents, use in labeling 150–151
Specular Reflection-FTIR 30
Spock ear tips 159
spot tests 27–29
stable isotope analysis 105–106
stereolithography (SLA) 10
Sterilite **13**, **95**, 98
Stone Paper 8
Styrofoam **13**, **95**, 102
sulfur 15, 46, 48, 110, 114, 115, **116**, 131
sunlight 21–22, 23, 69, 70–71, 74, 96
supports, evaluating 173
surveys: 9/11 Memorial & Museum 146–150; Cooper Hewitt 140–142; database 144–145; Deutsches

Bergbau-Museum Bochum, Leibniz-Museum für Geo-Ressourcen 142–146; Harvard Art Museums 77–78; objective descriptions 143–144; quantifying condition 146; tools for 142
synthetic paper 8–9
synthetic plastics 7 *see also* individual plastics

Teflon **13**, *43*, **95**, 103, 151
temperature 74–75; A-D strips and 119; CN conservation 51–52; evaluation of environment 173–174; Julia Child's kitchen 179–180; mitigation 76–79; monitors 75–76; sources 75
textiles: additives 17; handling of human-made fiber-based 130–131; labeling 151–152
thermohygrographs 75
thermoplastic plastics **13**, 14–15; semi-crystalline/amorphous 16
thermoset plastics **13**, 14
Time Weighted Preservation Index (TWPI) 157–158
timelines 42–44
trade names and patents 34–36
triphenyl phosphate (TPP) 20–21, 50, *84*, 85
Tupperware **13**, 35–36, **95**, 98, 160
Tyvek **13**, 94, **95**, 96, 97–98, 107, 130

U.S. National Park Service, reformatting CN film 165
UV light 70–71; blocking films/sleeves 74; photographic documentation 138; synthetic paper and 9; visual observation of degradation 169–170
UV stabilizers 17, 21

ventilation 19, 50, 51–52, 54, 65, 75, 90, 115, 131, 149, 157, 164, 166, 172, 180–181
vinegar 21, 53, 65, 117, 121–122; *see also* acetic acid
vinyl gloves 131
virtual preservation, 3D imaging 149
viscose 52
VoCA Artist Interview Workshops 132–134
Volara **13**, 94, **95**, 96–97
volatile organic compounds (VOCs) 18, 32, 64, 65, 83, 88, 98, 122, 131, 173
Vulcanite **14**, 46
vulcanized rubber *see* hard rubber (HR)

watches, storage impact on PVC 59–62
waterproofing 17

X-ray diffraction (XRD) 104
X-ray fluorescence (XRF) 30, 32, 58, 61, 104

yellowing/yellowed 21, *23*, *53*, *54*; Chanel suit 135–136; LDPE bags 99–100; papers 156; polypropylene (PP) 102; polystyrene (PS) 102; polyurethane foam 63; Shervone Neckles artwork 162; watches 60

zeolites 51, 58, 59, 90, 172